DISTRIBUTION THEORY OF
RUNS AND PATTERNS
AND ITS
APPLICATIONS

A Finite Markov Chain Imbedding Approach

DISTRIBUTION THEORY OF
RUNS AND PATTERNS
AND ITS
APPLICATIONS

A Finite Markov Chain Imbedding Approach

JAMES C. FU
University of Manitoba, Canada

W. Y. WENDY LOU
University of Toronto, Canada

World Scientific
New Jersey • London • Singapore • Hong Kong

Published by

World Scientific Publishing Co. Pte. Ltd.

5 Toh Tuck Link, Singapore 596224

USA office: Suite 202, 1060 Main Street, River Edge, NJ 07661

UK office: 57 Shelton Street, Covent Garden, London WC2H 9HE

Library of Congress Cataloging-in-Publication Data
Fu, James C.
    Distribution theory of runs and patterns and its applications / James C. Fu, W.Y. Wendy Lou.
       p. cm.
    Includes bibliographical references and index.
    ISBN 981-02-4587-4 (alk. paper)
       1. Markov processes. 2. Random variables. 3. Distribution (Probability theory). I. Lou,
W. Y. Wendy. II. Title.

QA274.7.F8 2003
519.2'33--dc21   2003053824

British Library Cataloguing-in-Publication Data
A catalogue record for this book is available from the British Library.

1003814936

Printed in Singapore.

To our parents

Preface

It is the purpose of this book to provide a rigorous, comprehensive introduction to the finite Markov chain imbedding technique for studying the distributions of runs and patterns from a unified and intuitive viewpoint, away from the lines of traditional combinatorics. Over the past two decades, considerably many new results related to the distributions of runs and patterns have been obtained through this approach.

The central theme of finite Markov chain imbedding, as the name suggests, is to properly imbed the random variables of interest into the framework of a finite Markov chain, and the resulting representations of the underlying distributions are compact and very amenable to further study of associated properties. In this book, the concept of finite Markov chain imbedding is systematically developed, and its utility is illustrated through practical applications to a variety of fields, including the reliability of engineering systems, hypothesis testing, quality control, and continuity measurement in the health care sector.

This book is restricted to discrete sample spaces, a restriction which serves to make this work accessible to a wider audience by simplifying the theoretical results and their applications. The runs and patterns considered herein are largely defined on sequences of Markov-dependent two- as well as multi-state trials with practical applications in mind; those defined on random permutations of integers, such as the Eulerian and Simon Newcomb numbers, are also treated using an additional insertion procedure. The content of this book is geared mainly towards researchers who are using the distribution theory of runs and patterns in various applied areas of statistics, probability and combinatorics, but it could also form the basis of a one-semester special-topics course at the fourth-year undergraduate or at the first-year graduate level.

We wish to acknowledge the assistance of Y. M. Chang and B. C. Johnson for proofreading early drafts of the book, as well as the encouragement from our colleagues at the University of Manitoba and the University of Toronto. We are also indebted to our families for their endless support. Lastly, we wish to thank Ms. E. H. Chionh of World Scientific Publishing Co. for her patience and managerial support.

JAMES C. FU

W.Y. WENDY LOU

Winnipeg, Manitoba
Toronto, Ontario

Contents

Chapter 1

Introduction

The occurrence of runs and patterns in a sequence of discrete trial outcomes or random permutations is an important concept in various areas of science, including reliability engineering, quality control, psychology, sociology, DNA sequence matching, and hypothesis testing. Results for the probability distributions of elementary runs and patterns were derived sporadically in the literature until about the 1940s, when a number of pioneering studies on more complex runs and patterns were published: for example, Wishart and Hirshfeld (1936), Cochran (1938), Mood (1940), Wald and Wolfowitz (1940), Mosteller (1941), and Wolfowitz (1943). Most of these studies focused on finding the conditional distribution of success runs given the total number of successes in a sequence of two-state trials. A recent book by Balakrishnan and Koutras (2002) provides an excellent, comprehensive review of historical and current developments in the distribution theory of runs and scan statistics.

Traditionally, the distributions of runs and patterns were studied via combinatorial analysis. For example, Mood (1940) wrote: "The distribution problem is, of course, a combinatorial one, and the whole development depends on some identities in combinatory analysis". However, finding the appropriate combinatorial identities to derive the probability distributions can be difficult, if not impossible, for complex runs and patterns, and this perhaps is the reason why the exact distributions of many common statistics defined on runs and patterns remain unknown. Furthermore, the required identities often differ even for similar runs and patterns, and hence, even in the simplest case of independent and identically distributed ($i.i.d.$) two-state trials (so-called "Bernoulli trials"), each new distribution

1

problem typically has to be studied on a case by case basis using the combinatorial approach. For example, only relatively recently did Philippou and Makri (1986) and Hirano (1986), independently and via combinatory analysis, obtain the exact distribution of the traditional runs statistic $N_{n,k}$, the number of non-overlapping consecutive k successes in a sequence of n Bernoulli trials:

$$P(N_{n,k} = x) = \sum_{m=0}^{k-1} \sum_{\substack{x_1+2x_2+\cdots+ \\ kx_k=n-m-kx}} \binom{x_1 + \cdots + x_k + x}{x_1, \cdots, x_k, x} p^n \left(\frac{q}{p}\right)^{x_1+\cdots+x_k},$$

for $x = 0, 1, \cdots, [n/k]$, with success and failure probabilities denoted by p and $q = 1 - p$, respectively.

Another approach for determining an exact probability distribution involves deriving the generating function $\varphi(s)$ for the non-negative integer random variable $X_n(\Lambda)$ associated with the pattern Λ (*e.g.* $X_n(\Lambda)$ could be the number of occurrences of the pattern Λ in n trials), and then differentiating $\varphi(s)$ x times to yield the probability distribution function (*pdf*) given by $P(X_n(\Lambda) = x)$; this approach was introduced by Feller (1968) using the theory of recurrent events. For example, the generating function of the waiting time $W(\Lambda)$, the number of Bernoulli trials until the first occurrence of the pattern Λ consisting of k consecutive successes, was given by Feller as

$$\varphi_W(s) = \frac{p^k s^k (1 - ps)}{1 - s + q p^k s^{k+1}}.$$

For more complex runs and patterns, the generating functions can be difficult to differentiate a large number of times, and approximation techniques may have to be employed. Feller used the method of partial fraction expansion, which can require efficient numerical methods for computing roots of polynomials.

Throughout this book, we will study distribution problems of runs and patterns from a, in our opinion, more unified and intuitive viewpoint, away from the lines of traditional combinatorics. The approach taken is to properly imbed the random variable $X_n(\Lambda)$ into a finite Markov chain $\{Y_t\}$, so that the probability of $X_n(\Lambda) = x$ can be expressed in terms of the probability of the Markov-chain outcome Y_n residing in a subset C_x of the state space; *i.e.*

$$P(X_n(\Lambda) = x) = P(Y_n \in C_x),$$

where the right-hand-side probability can be easily computed through the transition probability matrices of the Markov chain. This representation of the underlying distribution of $X_n(\Lambda)$ is compact, easy to compute, and quite amenable to further analysis. The method is largely dependent on being able to construct a proper Markov chain associated with the random variable $X_n(\Lambda)$, but once the chain is constructed, the linearity of the Markov chain reduces the computational complexity often associated with combinatorial and generating-function techniques for computing the exact *pdf*s of runs and patterns.

Early results in the distribution theory of runs and patterns were derived almost exclusively under the assumption of Bernoulli or *i.i.d.* multi-state trials. A great advantage of the finite Markov chain imbedding technique is that it can be applied not only to *i.i.d.* cases, but, with little additional effort, also to Markov-dependent multi-state trials, regardless of the counting procedures specified for overlapping patterns (overlap vs. non-overlap counting); it can also be extended to various types of runs and patterns on random permutations. Recently this method has been adopted by a number of researchers to study various distributions of runs and patterns: for example, Antzoulakos (1999, 2001), Boutsikas and Koutras (2000a,b), Doi and Yamamoto (1998), Fu (1985, 1986, 1996), Fu and Koutras (1994), Fu and Lou (2000a,b), Han and Aki (2000a,b), Johnson (2002), Koutras (1996a,b, 1997a,b, 2003), Koutras and Alexandrou (1995), Lou (1996, 2000, 2001), and Nishimura and Sibuya (1997). We will touch on some of these recent works, but our corresponding formulations may differ slightly in order to treat all problems using a common imbedding approach.

This book is not a review book for the theory of runs and patterns, nor is it intended to be used primarily as a course textbook; it is mainly aimed at researchers in applied statistics and probability who are interested in using the finite Markov chain imbedding technique to study the distributions of runs and patterns arising in specific applications. The contents of the book are largely based on recent developments in this area, but are presented in a manner that does not require knowledge of advanced concepts in mathematics or probability; a background in probability theory, at the level of, for example, Feller's (1968) book *"An Introduction to Probability Theory and Its Applications, Volume I"*, is assumed.

This book is organized in the following way. In Chapter 2, we introduce the basic ideas and techniques of finite Markov chain imbedding. This chapter lays the foundation for computing the *pdf*s of runs and patterns, including waiting-time distributions. Chapter 3 examines the distributions of runs and patterns associated with two-state trials, and in Chapter 4, the extension to multi-state trials via the forward and backward principle is treated. Chapter 5 mainly studies the waiting-time distributions of simple and compound patterns, as well as their generating functions and large deviation approximations. In Chapter 6, the finite Markov chain imbedding technique is extended to the study of distributions of patterns in random permutations of integers, focusing in detail on the Eulerian and the Simon Newcomb numbers. Chapter 7 covers several applications of the distribution theory of runs and patterns in the areas of the reliability of engineering systems, hypothesis testing, continuity measurement in health care, and quality control.

Chapter 2

Finite Markov Chain Imbedding

2.1 Finite Markov Chain

Let $\Omega = \{1, 2, \cdots, m\}$ $(m < \infty)$ be a finite state space, and let $\{Y_t\} = \{Y_0, Y_1, \cdots, Y_t, \cdots\}$ be a sequence of random variables defined on Ω.

Definition 2.1 The sequence $\{Y_t\}$ will be called a finite Markov chain if, for any sequence $\{Y_0 = i_0,\ Y_1 = i_1, \cdots, Y_{t-1} = i_{t-1},\ Y_t = i_t\}$, $t = 1, 2, \cdots$, we have

$$P(Y_t = i_t | Y_{t-1} = i_{t-1}, \cdots, Y_0 = i_0) = P(Y_t = i_t | Y_{t-1} = i_{t-1}). \qquad (2.1)$$

In other words, the sequence is a Markov chain if the probability that the system enters the state i_t at time t depends only on the immediately preceding state i_{t-1} at time $t - 1$. Or more succinctly, viewed from the state at time $t - 1$, the future is independent of the past.

The conditional probabilities

$$P(Y_t = j | Y_{t-1} = i) \equiv p_{ij}(t), \qquad (2.2)$$

$i, j \in \Omega$, are called one-step transition probabilities for the system at time t. The transition probabilities $p_{ij}(t)$, $1 \le i, j \le m$, may be represented as an $m \times m$ matrix

$$\boldsymbol{M}_t = \big(p_{ij}(t)\big) = \begin{pmatrix} p_{11}(t) & p_{12}(t) & \cdots & p_{1m}(t) \\ p_{21}(t) & \cdots & \cdots & \cdots \\ \vdots & \ddots & \ddots & \cdots \\ p_{m1}(t) & p_{m2}(t) & \cdots & p_{mm}(t) \end{pmatrix}_{m \times m}.$$

The matrices M_t, $t = 1, 2, \cdots$, are called one-step transition probability matrices.

Definition 2.2 A Markov chain $\{Y_0, Y_1, \cdots\}$ is *homogeneous* if the transition probabilities are constant in time, *i.e.* $P(Y_t = j | Y_{t-1} = i) = p_{ij}$ for any $i, j \in \Omega$, and all $t = 1, 2, \cdots$.

This definition is equivalent to saying that the transition probability matrices of a homogeneous Markov chain may be represented by the single matrix

$$M = M_t = (p_{ij}), \qquad \text{for all } t = 1, 2, \cdots,$$

where the transition probabilities p_{ij} are independent of the time index t.

The set of probabilities at time 0, $P(Y_0 = i)$ for $i = 1, \cdots, m$, is referred to as the *initial distribution* of the Markov chain. Given an initial distribution and the transition probabilities of a Markov chain, the *joint distribution* of the chain can be computed as follows:

$$
\begin{aligned}
P(Y_n = i_n, \cdots, Y_1 = i_1, Y_0 = i_0) &= P(Y_n = i_n | Y_{n-1} = i_{n-1}) \cdots \\
&\quad \cdots P(Y_1 = i_1 | Y_0 = i_0) P(Y_0 = i_0).
\end{aligned}
$$

$$(2.3)$$

Markov chains have been used in the modeling of a vast number of applications. Here we give two simple examples that are often seen in applied probability theory:

Example 2.1 (Gambler's ruin problem). Consider a gambler who wins and loses a dollar with probabilities p and $q = 1 - p$, respectively. Suppose the gambler has an initial capital of a dollars. The gambler quits playing the game either when he is out of capital ("ruined") or when he attains a fortune of $a + b$ dollars (with net gain $b > 0$).

The sequence of the gambler's amount of capital, $\{Y_t, t = 0, 1, 2, \cdots\}$, forms a homogeneous Markov chain with state space $\Omega = \{0, 1, 2, \cdots, a, a + 1, \cdots, a + b\}$ and the following transition probabilities:

$$
p_{ij} = \begin{cases} p & \text{if } j = i + 1 \\ q & \text{if } j = i - 1, \end{cases}
$$

for $i = 1, 2, \cdots, a + b - 1$, $p_{00} = p_{a+b,a+b} = 1$, and zero elsewhere. The states 0 and $a + b$ are referred to as absorbing states since, once entered, they will never be exited. The Markov chain has the transition probability matrix

$$
M = \begin{array}{c} 0 \\ 1 \\ \vdots \\ a-1 \\ a \\ a+1 \\ \vdots \\ a+b \end{array}
\left(\begin{array}{ccccccccc}
1 & 0 & 0 & 0 & & & & & \\
q & 0 & p & 0 & & & \boldsymbol{O} & & \\
& \ddots & \ddots & \ddots & & & & & \\
& & \ddots & \ddots & \ddots & & & & \\
& & & q & 0 & p & & & \\
& & & & \ddots & \ddots & \ddots & & \\
& \boldsymbol{O} & & & & q & 0 & p \\
& & & & & 0 & 0 & 1
\end{array}\right),
$$

where $\boldsymbol{O}$ represents a zero matrix, and the chain has the initial distribution $P(Y_0 = a) = 1$. $\diamond$

Example 2.2 (Randomly inserting balls into urns). Consider a sequence of independent trials, each consisting of inserting a ball at random into one of k urns. We say the system $\{Y_t : t = 0, 1, \cdots\}$ is in state i if exactly i urns are occupied. This system forms a Markov chain on the state space $\Omega = \{0, 1, \cdots, k\}$ with transition probabilities

$$
p_{ij} = \begin{cases}
\frac{i}{k} & \text{if } j = i \\
\frac{k-i}{k} & \text{if } j = i+1 \\
0 & \text{otherwise,}
\end{cases}
$$

for $i = 0, 1, \cdots, k$, and initial distribution $P(Y_0 = 0) = 1$. The transition probability matrix is given by

$$
M = \begin{array}{c} 0 \\ 1 \\ \vdots \\ i \\ \vdots \\ k-1 \\ k \end{array}
\left(\begin{array}{ccccccc}
0 & 1 & 0 & & & & \\
0 & \frac{1}{k} & \frac{k-1}{k} & & & \boldsymbol{O} & \\
& & \ddots & \ddots & & & \\
& & & \frac{i}{k} & \frac{k-i}{k} & & \\
& & & & \ddots & \ddots & \\
& \boldsymbol{O} & & & & \frac{k-1}{k} & \frac{1}{k} \\
& & & & & 0 & 1
\end{array}\right).
$$

$\diamond$

Further examples of this type may be found in, for example, Feller (1968) and Ross (2000). Of course, there will also be many more examples of Markov chains in later sections of this book.

2.2   Chapman-Kolmogorov Equation

For a non-homogeneous Markov chain $\{Y_t\}$, the n-step transition probabilities $P(Y_t = j | Y_{t-n} = i) = p_{ij}^{(n)}(t)$ can be obtained from the one-step transition probabilities by an important identity known as the Chapman-Kolmogorov equation. If $n = 2$, we have, for $t \geq 2$,

$$p_{ij}^{(2)}(t) = \sum_k P(Y_{t-1} = k | Y_{t-2} = i) P(Y_t = j | Y_{t-1} = k) = \sum_k p_{ik}(t-1) p_{kj}(t),$$
(2.4)

which corresponds to summing over all possible intermediate states k in the transition from state i to state j.

If $\{Y_t\}$ is a homogeneous Markov chain, then Eq. (2.4) yields the two-step ($n = 2$) transition probabilities

$$p_{ij}^{(2)} = \sum_k P(Y_{t-1} = k | Y_{t-2} = i) P(Y_t = j | Y_{t-1} = k) = \sum_k p_{ik} p_{kj}, \quad (2.5)$$

which are independent of t. Hence, from Eq. (2.5), the two-step transition probability matrix $\boldsymbol{M}^{(2)} = (p_{ij}^{(2)})$ satisfies the identity $\boldsymbol{M}^{(2)} = \boldsymbol{M}^2$. Similarly, for the n-step transition probabilities of a homogeneous Markov chain, the Chapman-Kolmogorov identity,

$$p_{ij}^{(n)} = \sum_k p_{ik}^{(s)} p_{kj}^{(n-s)},$$
(2.6)

holds for every intermediate step $s = 1, \cdots, n-1$. It follows from Eqs. (2.5) and (2.6) that

$$\boldsymbol{M}^{(n)} = \boldsymbol{M}^{(s)} \boldsymbol{M}^{(n-s)} = \boldsymbol{M}^n.$$
(2.7)

For a homogeneous Markov chain $\{Y_t\}$, and any subset C of the state space Ω, it follows from Eq. (2.7) that the conditional probability of the system state Y_n residing in C at time index n, given the initial distribution $\boldsymbol{\xi}_0 = (P(Y_0 = 1), \cdots, P(Y_0 = m))$, is

$$P(Y_n \in C | \boldsymbol{\xi}_0) = \boldsymbol{\xi}_0 \boldsymbol{M}^n \boldsymbol{U}^{'}(C),$$
(2.8)

where $U'(C)$ denotes the transpose of $U(C)$, $U(C) = \sum_{i \in C} e_i$, and $e_i = (0, \cdots, 1, 0, \cdots, 0)_{1 \times m}$ is a unit row vector with 1 corresponding to the i-th state and zero elsewhere. More generally, if $\{Y_t\}$ is a non-homogeneous Markov chain, it can be shown (*e.g.* Feller 1968) that the conditional probability of $Y_n \in C$ given $\boldsymbol{\xi}_0$ can be simply represented as

$$P(Y_n \in C | \boldsymbol{\xi}_0) = \boldsymbol{\xi}_0 (\prod_{t=1}^{n} M_t) U'(C). \tag{2.9}$$

Equations (2.8) and (2.9) are two indispensable tools for evaluating probabilities of various events associated with homogeneous and non-homogeneous Markov chains, respectively.

2.3   Tree-Structured Markov Chain

In order to broaden the possible applications, it is useful to consider a simple extension of the preceding methodology to Markov chains defined on state spaces of different sizes. Let $\{Y_t\}$ be a sequence of random variables defined on a sequence of state spaces $\{\Omega_t\}$, respectively. The sequence $\{Y_t\}$ is referred to as a tree-structured Markov chain if $\{Y_t\}$ is a Markov chain with transition matrices

$$M_t = (p_{ij}(t)), \qquad t = 1, 2, \cdots,$$

where, for $i \in \Omega_{t-1}$ and $j \in \Omega_t$,

$$p_{ij}(t) = P(Y_t = j | Y_{t-1} = i).$$

Note that since the state spaces in the sequence $\{\Omega_t\}$ may have different sizes, the transition matrices M_t may be rectangular instead of square; *i.e.* M_t, $t = 1, 2, \cdots$, are $card(\Omega_{t-1}) \times card(\Omega_t)$ matrices, where $card(\Omega_t)$ stands for the cardinal number of the state space Ω_t. The sequence of transition probability matrices $\{M_t\}$ still determines the tree-structured Markov chain $\{Y_t\}$, and the Chapman-Kolmogorov equation remains applicable.

For any subset $C \subseteq \Omega_n$, the conditional probability of $Y_n \in C$ given the initial distribution $\boldsymbol{\xi}_0$ can be evaluated via

$$P(Y_n \in C | \boldsymbol{\xi}_0) = \boldsymbol{\xi}_0 (\prod_{t=1}^{n} M_t) U'_n(C), \tag{2.10}$$

where $U_n(C) = \sum_{i \in C} e_i$ and $e_i = (0, \cdots, 0, 1, 0, \cdots, 0)$ is a $1 \times card(\Omega_n)$ unit row vector with 1 at the location associated with state i. If all the state spaces are identical ($\Omega_1 = \cdots = \Omega_n$), then Eq. (2.10) reduces to Eq. (2.9).

2.4   Runs and Patterns

Traditionally, within a sequence of Bernoulli (*i.i.d.* success-failure) trials, a run denotes a sequence of consecutive successes or failures. For example, a success run of size 4 implies the pattern $SSSS$. Several runs statistics that are often used in statistics and applied probability for a sequence of n Bernoulli trials are:

(i) $N_{n,k}$, the number of non-overlapping consecutive k successes, in the sense of Feller's (1968) counting;

(ii) $M_{n,k}$, the number of overlapping consecutive k successes;

(iii) $E_{n,k}$, the number of success runs of size exactly k, in the sense of Mood's (1940) counting;

(iv) $G_{n,k}$, the number of success runs of size greater than or equal to k; and

(v) $L_n(S)$, the size of the longest success run.

Perhaps the easiest way to understand the precise definitions of these runs statistics and the overlap/non-overlap counting procedure is by means of the following example. Suppose that there are $n = 10$ Bernoulli trials, with realization $SSFSSSSFFF$. Then $L_{10}(S) = 4$, and for $k = 2$, we have $N_{10,2} = 3$, $M_{10,2} = 4$, $E_{10,2} = 1$ and $G_{10,2} = 2$.

From the definitions of these runs statistics, it follows by inspection that the following relationships are always true:

$$E_{n,k} \leq G_{n,k} \leq N_{n,k} \leq M_{n,k},$$

$$E_{n,k} = G_{n,k} - G_{n,k+1},$$

$$L_n(S) < k \text{ if and only if } N_{n,k} = 0.$$

To extend the definitions of runs, let us consider a sequence $\{X_t\}_{t=1}^n$ of n multi-state trials, each of which has $m \geq 2$ states or symbols as possible outcomes. These symbols are denoted by $b_1, b_2, \cdots, b_m$, and occur with probabilities $p_1, p_2, \cdots, p_m$, respectively. We then define three types of general patterns: a simple pattern, a compound pattern, and a series pattern.

Definition 2.3 Λ is a *simple pattern* if Λ is composed of a specified sequence of k symbols, *i.e.* $\Lambda = b_{i_1} \cdots b_{i_k}$, with $i_j \in \{1, \cdots, m\}$ for all $j = 1, \cdots, k$. The length of the pattern is fixed, and the symbols in the pattern may be repeated.

Success runs and failure runs of size k are thus simple patterns under this definition, and in fact any fixed-length sequence of successes and failures, say $\Lambda = SSFSF$, can be considered as a simple pattern within a sequence of n two-state ($m = 2$) trials.

Next, let Λ_1 and Λ_2 be two simple patterns of lengths/sizes k_1 and k_2, respectively. We say that Λ_1 and Λ_2 are *distinct* if neither Λ_1 belongs to Λ_2 nor Λ_2 belongs to Λ_1. We define $\Lambda_1 \cup \Lambda_2$ to denote the occurrence of either pattern Λ_1 or pattern Λ_2, and define $\Lambda_1 * \Lambda_2$ to denote the occurrence of pattern Λ_1 followed by pattern Λ_2 (with possibly a gap between them).

Definition 2.4 Λ is a *compound pattern* if it is the union of $1 < l < \infty$ overlapping/non-overlapping distinct simple patterns, *i.e.* $\Lambda = \cup_{i=1}^{l}\Lambda_i$.

Definition 2.5 Λ is a *series pattern* if Λ is composed of an ordered sequence of $1 < l < \infty$ non-overlapping distinct simple patterns Λ_i, *i.e.* $\Lambda = \Lambda_1 * \Lambda_2 * \cdots * \Lambda_l$.

Throughout this book, the random variable $X_n(\Lambda)$ represents the number of occurrences of the pattern Λ in a sequence of n multi-state trials, using either overlap or non-overlap counting. In order to clarify the three pattern definitions and the two types of counting conventions for multi-state trials, we give the following example.

Example 2.3 Let $\{X_t\}_{t=1}^{16}$ be a sequence of sixteen four-state trials, where the possible outcomes for each trial are A, C, G, and T. Let $\Lambda_1 = AGAG$ and $\Lambda_2 = AGT$ be two distinct simple patterns, $\Lambda = \Lambda_1 \cup \Lambda_2$ be a compound pattern, and $\Lambda^* = \Lambda_1 * \Lambda_2$ be a series pattern. Suppose the realization of this sequence of sixteen trials is $TAGAGAGTCAGAGTCC$, then

(i) $X_{16}(\Lambda_1)$ equals 3 with overlap counting and equals 2 with non-overlap counting,

(ii) $X_{16}(\Lambda)$ equals 5 with overlap counting and equals 3 with non-overlap counting, and

(iii) $X_{16}(\Lambda^*)$ equals one.   ◇

The above definitions of runs and patterns in a sequence of multi-state trials can also be extended to random permutations $\{\pi : \pi = (\pi(1),\cdots,\pi(n))\}$ of n integers $\{1, 2, \cdots, n\}$. For example, the Eulerian number $E(\pi, n)$, the number of rises in a random permutation π (see Carlitz 1964, Tanny 1973, and Worpitzky 1883), could be viewed as a random variable $X_n(\Lambda)$ with the pattern Λ being a rise. Mathematically, the Eulerian number can be defined as

$$E(\pi, n) = X_n(\Lambda) = \sum_{i=0}^{n-1} I(\pi, i),$$

where

$$I(\pi, i) = \begin{cases} 1 & \text{if } \pi(i) < \pi(i+1) \\ 0 & \text{otherwise,} \end{cases}$$

for $i = 1, \cdots, n - 1$, with $I(\pi, 0) = 1$ by convention (the starting "gap" preceding the first permutation is always considered a rise). For example, the number of rises $E(\pi, 9)$ in the random permutation $\pi = (321459768)$ of 9 integers is 5.

In view of the above definitions and examples, one should expect the exact distribution of the random variable $X_n(\Lambda)$ to depend heavily on three important factors: (a) the structure of the pattern Λ, (b) the structure of the sequence $\{X_t\}_{t=1}^n$ of n trials (or random permutations), and (c) the counting procedure (overlap or non-overlap counting). Due to these factors, the analytical determination of exact distributions via traditional approaches such as combinatorics can be quite challenging and is generally complex, involving special identities and lengthy algebra. Consequently, the exact distributions of many statistics used in practical applications have never been studied using such methods, especially when the underlying trials are non-*i.i.d.* (*e.g.* Markov-dependent).

In the following subsection, we describe a technique that allows one to obtain a compact matrix representation for the exact distribution in a relatively simple and universal manner by imbedding the random variable $X_n(\Lambda)$ into a finite Markov chain; the resulting expression is also very amenable to further analysis of statistical properties, to the development of large deviation approximations, and to efficient numerical implementation for the computation of exact probabilities.

2.5 Finite Markov Chain Imbedding

The finite Markov chain imbedding technique for finding the distribution of the random variable $X_n(\Lambda)$ has its early origins in a series of papers by Fu (1985, 1986), Fu and Hu (1987), Chao and Fu (1989, 1991), and Fu and Lou (1991). The term "finite Markov chain imbeddable" to describe a random variable was formally introduced by Fu and Koutras (1994).

Let $\Gamma_n = \{0, 1, \cdots, n\}$ be an index set, and let $\Omega = \{a_1, a_2, \cdots, a_m\}$ be a finite state space.

Definition 2.6 The non-negative integer-valued random variable $X_n(\Lambda)$ is *finite Markov chain imbeddable* if:

(a) there exists a finite Markov chain $\{Y_t : t \in \Gamma_n\}$ defined on a finite state space Ω with initial probability vector $\boldsymbol{\xi}_0$,
(b) there exists a finite partition $\{C_x : x = 0, 1, \cdots, l_n\}$ on the state space Ω, and
(c) for every $x = 0, 1, \cdots, l_n$, we have

$$P(X_n(\Lambda) = x) = P(Y_n \in C_x | \boldsymbol{\xi}_0).$$

Let $\{M_t\}_{t=1}^n$ be the sequence of $m \times m$ transition probability matrices of the finite Markov chain $\{Y_t\}$ defined on the state space Ω with initial probability distribution $\boldsymbol{\xi}_0 = (P(Y_0 = a_1), P(Y_0 = a_2), \cdots, P(Y_0 = a_m))$.

Theorem 2.1 *If $X_n(\Lambda)$ is finite Markov chain imbeddable, then*

$$P(X_n(\Lambda) = x) = \boldsymbol{\xi}_0 (\prod_{t=1}^n M_t) U'(C_x), \qquad (2.11)$$

where $U(C_x) = \sum_{r:a_r \in C_x} e_r$, e_r is a $1 \times m$ unit row vector corresponding to state a_r, $\boldsymbol{\xi}_0$ is the initial probability vector, and M_t, $t = 1, \cdots, n$, are the transition probability matrices of the imbedded Markov chain.

Proof. Since $X_n(\Lambda)$ is finite Markov chain imbeddable, it follows from Definition 2.6(a) that there exists a finite Markov chain $\{Y_t : t \in \Gamma_n\}$ with initial probability $\boldsymbol{\xi}_0$. By the Chapman-Kolmogorov equation described in Section 2.2, for every $a_r \in \Omega$, we have

$$P(Y_n = a_r | \boldsymbol{\xi}_0) = \boldsymbol{\xi}_0 (\prod_{t=1}^n M_t) e_r'.$$

Furthermore, it follows from Definitions 2.6(b) and (c) that, for every $x = 0, 1, \cdots, l_n$,

$$
\begin{aligned}
P(X_n(\Lambda) = x) &= P(Y_n \in C_x | \boldsymbol{\xi}_0) = \sum_{a_r \in C_x} P(Y_n = a_r | \boldsymbol{\xi}_0) \\
&= \sum_{a_r \in C_x} \boldsymbol{\xi}_0 (\prod_{t=1}^{n} \boldsymbol{M}_t) \boldsymbol{e}_r' \\
&= \boldsymbol{\xi}_0 (\prod_{t=1}^{n} \boldsymbol{M}_t) \boldsymbol{U}'(C_x).
\end{aligned}
$$

□

The k-th moment $E(X_n^k(\Lambda))$, $k = 1, 2, \cdots$, can be written as

$$
E(X_n^k(\Lambda)) = \boldsymbol{\xi}_0 (\prod_{t=1}^{n} \boldsymbol{M}_t) \boldsymbol{V}_k', \tag{2.12}
$$

where

$$
\boldsymbol{V}_k = \sum_{x=0}^{l_n} x^k \boldsymbol{U}(C_x).
$$

Similarly, the probability generating function for the random variable $X_n(\Lambda)$ can be written as

$$
\varphi(s) = \boldsymbol{\xi}_0 (\prod_{t=1}^{n} \boldsymbol{M}_t) \boldsymbol{W}'(s), \tag{2.13}
$$

where

$$
\boldsymbol{W}(s) = \sum_{x=0}^{l_n} s^x \boldsymbol{U}(C_x).
$$

Moments and probability generating functions will be discussed within the context of specific applications in subsequent sections.

Example 2.4 (Number of pairs of identical successive outcomes). Let $\{X_t : t = 0, 1, \cdots, n\}$ be a sequence of homogeneous Markov-dependent m-state trials with transition probability matrix $\boldsymbol{A}_{m \times m} = (p_{ij})$ and initial probability distribution $\boldsymbol{\pi}_0 = (P(X_0 = 1), \cdots, P(X_0 = m)) = (1/m, 1/m, \cdots, 1/m)$. Define a sequence of indicator functions

$$
I_t = \begin{cases} 1 & \text{if } X_t = X_{t-1} \\ 0 & \text{otherwise,} \end{cases}
$$

for $t = 1, \cdots, n$.

In this example, we are interested in the number of times that a particular outcome (one of m possible outcomes) at a given trial is repeated at the immediately following trial. In mathematical terms, we define the pattern Λ to denote such a repeated outcome, a pattern which is present at time index $1 \le t \le n$ if $X_{t-1} = X_t$ or, equivalently in terms of the above indicator function, if $I_t = 1$. The runs statistic

$$X_n(\Lambda) = \sum_{t=1}^{n} I_t$$

then corresponds to the number of times that the pattern Λ occurred in the sequence of Markov-dependent m-state trials $\{X_t\}_{t=0}^{n}$. In the healthcare sector, for example, the statistic $X_n(\Lambda)/n$ is known as $SECON$, and forms the primary measure of sequential continuity in a series of n patient visits to m possible healthcare providers (Steinwachs 1979).

One difficulty here is that the random variables $\{I_t\}$ are not independent and do not together form a Markov chain, even if the sequence $\{X_t\}_{t=0}^{n}$ were drawn from *i.i.d.* m-state trials. In fact, one can show that the random variables $\{I_t\}$ are dependent and positively correlated by proving that $Cov(I_i, I_j) > 0$ for all i and j, with $Cov(I_i, I_j) \to 0$ as $|i - j| \to \infty$. However, as given below, the exact distribution can still be readily obtained using the finite Markov chain imbedding technique.

First, we decompose the transition probability matrix $\boldsymbol{A}$ into two matrices $\boldsymbol{G}$ and $\boldsymbol{D}$, where the latter matrix contains only the diagonal elements of $\boldsymbol{A}$; *i.e.*

$$\boldsymbol{A}_{m \times m} = \boldsymbol{G}_{m \times m} + \boldsymbol{D}_{m \times m},$$

where

$$\boldsymbol{G}_{m \times m} = \begin{pmatrix} 0 & p_{ij} & \cdots \\ & \ddots & \\ \cdots & p_{ij} & 0 \end{pmatrix} \text{ and } \boldsymbol{D}_{m \times m} = \begin{pmatrix} p_{11} & & \boldsymbol{O} \\ & \ddots & \\ \boldsymbol{O} & & p_{mm} \end{pmatrix}.$$

Let $\Omega = \{(u, v) : u = 0, \cdots, n, \text{and } v = 1, 2, \cdots, m\}$ be the state space containing a total of $(n+1)m$ states. Given n, define a finite homogeneous Markov chain $\{Y_t : t \in \Gamma_n\}$ on the state space Ω as

$$Y_t = \begin{cases} (\sum_{i=1}^{t} I_i, X_t) & 1 \le t \le n \\ (0, X_0) & t = 0, \end{cases}$$

with transition probability matrix

$$
M = \begin{pmatrix}
G & D & O & & & O \\
O & G & D & & & \\
 & \ddots & \ddots & \ddots & & \\
 & & \cdots & & G & D \\
O & & & & O & I
\end{pmatrix},
$$

where M is an $(n+1)m \times (n+1)m$ matrix, O represents the $m \times m$ zero matrix, and I is the $m \times m$ identity matrix. The states in M are arranged in lexicographical (dictionary) order. Lastly, define the partition $\{C_x : x = 0, 1, 2, \cdots, n\}$ on the state space Ω as

$$
C_x = \{(x, v) : v = 1, 2, \cdots, m\}.
$$

Given the above definitions for the Markov chain $\{Y_t\}$, the random variable $X_n(\Lambda)$ is, by Definition 2.6, finite Markov chain imbeddable and its exact distribution follows from Theorem 2.1: for $0 \le x \le n$,

$$
P(X_n(\Lambda) = x) = \boldsymbol{\xi}_0 \begin{pmatrix}
G & D & O & & & O \\
O & G & D & & & \\
 & \ddots & \ddots & \ddots & & \\
 & & \cdots & & G & D \\
O & & & & O & I
\end{pmatrix}^n U'(C_x), \qquad (2.14)
$$

where $\boldsymbol{\xi}_0 = (\boldsymbol{\pi}_0, 0, \cdots, 0)_{1 \times (n+1)m}$ is the initial distribution of the state vector Y_0, and $U(C_x) = (0, \cdots, 0, 1, 1, \cdots, 1, 0, \cdots, 0)$ is a $1 \times (n+1)m$ row vector with 1 at the coordinates associated with states in C_x and zero elsewhere. Further details and a numerical example for this problem will be given in Chapter 7. $\qquad \diamond$

Koutras and Alexandrou (1995) introduced the notion of finite Markov chain imbeddable variables of binomial type (MVBs), and many common statistics for runs and patterns fall into this special category. Let the partition $\{C_x\} = \{[(x, v) : v = 1, \cdots, r], \text{ for } x = 0, 1, \cdots, l_n\}$ be the partition of the state space Ω.

Definition 2.7 A random variable $X_n(\Lambda)$ is finite Markov chain imbeddable of binomial type if

(i) $X_n(\Lambda)$ is finite Markov chain imbeddable as in Definition 2.6, and
(ii) $P(Y_t = (y, j) | Y_{t-1} = (x, i)) \equiv 0$ for all $y \neq x$ or $x + 1$.

For any MVB, introduce two $r \times r$ transition probability matrices:

$$\boldsymbol{A}_t(x) = (a_{ij}(t)) = (P(Y_t = (x,j)|Y_{t-1} = (x,i)))$$

and

$$\boldsymbol{B}_t(x) = (b_{ij}(t)) = (P(Y_t = (x+1,j)|Y_{t-1} = (x,i)))\,.$$

Then the transition probability matrices $\boldsymbol{M}_t$ of the imbedded Markov chain have the following form:

$$\boldsymbol{M}_t = \begin{pmatrix} \boldsymbol{A}_t(0) & \boldsymbol{B}_t(0) & & & & \\ & \boldsymbol{A}_t(1) & \boldsymbol{B}_t(1) & & \boldsymbol{O} & \\ & & \ddots & & \ddots & \\ & & & \ddots & & \ddots \\ & \boldsymbol{O} & & & \boldsymbol{A}_t(l_n-1) & \boldsymbol{B}_t(l_n-1) \\ & & & & & \boldsymbol{A}_t(l_n) \end{pmatrix}, \quad (2.15)$$

for $t = 1, \cdots, n$, where the states are arranged in lexicographical (dictionary) order. There are many statistics for runs and patterns with transition matrices that are of this form, such as, for example, the runs statistics $N_{n,k}$, $M_{n,k}$, and $G_{n,k}$ introduced in Section 2.4 (and studied further in Chapter 3).

For MVBs, an efficient recursive equation for the distribution of $X_n(\Lambda)$ may be derived, one which partially takes advantage of the banded structure of the transition probability matrices $\boldsymbol{M}_t$. Let the row vector $\boldsymbol{\alpha}_t(x) = (P(Y_t = (x,1)), \cdots, P(Y_t = (x,r)))$, for $t = 1, \cdots, n$, so that the probability of $X_n(\Lambda) = x$ can be represented as

$$P(X_n(\Lambda) = x|\boldsymbol{\xi}_0) = \boldsymbol{\alpha}_n(x)\boldsymbol{1}', \text{ for all } x = 0, 1, \cdots, l_n, \quad (2.16)$$

where $\boldsymbol{1}' = (1, \cdots, 1)'$. Decompose $\boldsymbol{M}_t$ as $\boldsymbol{M}_t = \boldsymbol{K}_t + \boldsymbol{H}_t$, where $\boldsymbol{K}_t$ is a diagonal matrix with components $\boldsymbol{A}_t(x)$, for $x = 0, 1, \cdots, l_n$, and $\boldsymbol{H}_t$ is an upper-diagonal matrix with components $\boldsymbol{B}_t(x)$, for $x = 0, 1, \cdots,$ $l_n - 1$. From backward multiplication, $\boldsymbol{\xi}_0(\prod_{j=1}^t \boldsymbol{M}_j) = \boldsymbol{\xi}_0(\prod_{j=1}^{t-1} \boldsymbol{M}_j)\boldsymbol{M}_t$ $= \boldsymbol{\xi}_0(\prod_{j=1}^{t-1} \boldsymbol{M}_j)(\boldsymbol{K}_t + \boldsymbol{H}_t)$, it can be shown that the following recursive equations hold:

$$\begin{aligned} \boldsymbol{\alpha}_t(0) &= \boldsymbol{\alpha}_{t-1}(0)\boldsymbol{A}_t(0) & (2.17) \\ \boldsymbol{\alpha}_t(x) &= \boldsymbol{\alpha}_{t-1}(x-1)\boldsymbol{B}_{t-1}(x-1) + \boldsymbol{\alpha}_{t-1}(x)\boldsymbol{A}_t(x), \ x = 1, \cdots, l_n. \end{aligned}$$

Equation (2.17) provides an efficient algorithm for computing the probabilities $P(X_n(\Lambda) = x|\boldsymbol{\xi}_0) = \boldsymbol{\alpha}_n(x)\mathbf{1}'$, for all $x = 0, 1, \cdots, l_n$, and this is especially important when the dimension of the transition matrices $\boldsymbol{M}_t$ is large and the computational effort in naively calculating $\boldsymbol{\xi}_0(\prod_{t=1}^{n} \boldsymbol{M}_t)\boldsymbol{U}'(C_x)$ becomes prohibitive. From backward multiplication, the finite Markov chain imbedding technique often provides a recursive equation in a form similar to Eq. (2.17), a form which cannot, in general, be so easily obtained through the traditional combinatorial or renewal methods.

2.6   Absorbing State

In this section, some useful expressions are derived for the probability of entering an absorbing state. For clarity of the exposition, the focus will be on homogeneous Markov chains, but the ideas may be readily generalized to non-homogeneous cases.

A state $\alpha \in \Omega$ is called an absorbing state if, once the system enters state α, it never leaves; *i.e.* $p_{\alpha\alpha} \equiv 1$ (and $p_{\alpha b} \equiv 0$ for any $b \neq \alpha$). Let $A = \{\alpha_1, \cdots, \alpha_k\}$ be the set of all absorbing states of a homogeneous Markov chain $\{Y_t\}$ with transition probability matrix $\boldsymbol{M}$. Under appropriate arrangement of the state space, the transition probability matrix $\boldsymbol{M}$ can always be written in the following form:

$$\boldsymbol{M} = \left(\begin{array}{c|c} \boldsymbol{N}_{(m-k)\times(m-k)} & \boldsymbol{C}_{(m-k)\times k} \\ \hline \boldsymbol{O}_{k\times(m-k)} & \boldsymbol{I}_{k\times k} \end{array} \right), \tag{2.18}$$

where m and k $(m > k)$ are the numbers of states in Ω and A, respectively. The matrix $\boldsymbol{N}$ defined by Eq. (2.18) is referred to as the essential transition probability submatrix of the Markov chain. It plays an important role in studying the exact distributions of Markov-chain-imbeddable random variables, especially for associated distributions of waiting times.

Let $\boldsymbol{\xi}_0 = (\boldsymbol{\xi} : \mathbf{0})_{1\times m}$ be the initial distribution, where $\boldsymbol{\xi}=(\xi_1, \cdots, \xi_{m-k})$, $\mathbf{0} = (0, \cdots, 0)_{1\times k}$, and $\sum_{i=1}^{m-k} \xi_i = 1$, and let $(\mathbf{1} : \mathbf{0})_{1\times m}$ be a row vector, where $\mathbf{1} = (1, \cdots, 1)_{1\times(m-k)}$. The reason why we assume that the initial distribution has the form $(\boldsymbol{\xi} : \mathbf{0})$ is strictly for practical reasons, as most systems always start in a non-absorbing state.

Theorem 2.2   *Given a transition probability matrix $\boldsymbol{M}$ of a homogeneous Markov chain $\{Y_t\}$ in the form of Eq. (2.18), the probability for the time index n when the system first enters the set of absorbing states can be* *obtained from*

$$P(Y_n \in A, Y_{n-1} \notin A, \cdots, Y_1 \notin A | \boldsymbol{\xi}_0) = \boldsymbol{\xi} \boldsymbol{N}^{n-1}(\boldsymbol{I} - \boldsymbol{N})\boldsymbol{1}'. \qquad (2.19)$$

Proof. Since $\boldsymbol{M}$ has the form of Eq. (2.18), it follows that

$$\boldsymbol{M}^{n-1} = \left(\begin{array}{c|c} \boldsymbol{N}^{n-1} & \boldsymbol{K}_{n-1} \\ \hline \boldsymbol{O} & \boldsymbol{I} \end{array} \right), \qquad (2.20)$$

where $\boldsymbol{K}_{n-1} = (\boldsymbol{I} + \boldsymbol{N} + \cdots + \boldsymbol{N}^{n-2})\boldsymbol{C}$. Also, as all the states in A are absorbing states, it follows from the Chapman-Kolmogorov equation that

$$\begin{aligned} P(Y_{n-1} \notin A, \cdots, Y_1 \notin A | \boldsymbol{\xi}_0) &= P(Y_{n-1} \in \Omega - A | \boldsymbol{\xi}_0) \\ &= (\boldsymbol{\xi} : \boldsymbol{0})\boldsymbol{M}^{n-1}(\boldsymbol{1} : \boldsymbol{0})'. \qquad (2.21) \end{aligned}$$

Equation (2.19) may then be deduced using Eqs. (2.20) and (2.21) via

$$\begin{aligned} &P(Y_n \in A, Y_{n-1} \notin A, \cdots, Y_1 \notin A | \boldsymbol{\xi}_0) \\ &= P(Y_{n-1} \notin A, \cdots, Y_1 \notin A | \boldsymbol{\xi}_0) - P(Y_n \notin A, Y_{n-1} \notin A, \cdots, Y_1 \notin A | \boldsymbol{\xi}_0) \\ &= (\boldsymbol{\xi} : \boldsymbol{0})\boldsymbol{M}^{n-1}(\boldsymbol{I} - \boldsymbol{M})(\boldsymbol{1} : \boldsymbol{0})' = \boldsymbol{\xi} \boldsymbol{N}^{n-1}(\boldsymbol{I} - \boldsymbol{N})\boldsymbol{1}'. \end{aligned}$$

$\square$

In view of Eqs. (2.20) and (2.21), the following theorems are immediate consequences.

Theorem 2.3 *For any state $i \in \Omega - A$,*

$$P(Y_{n-1} = i, Y_{n-2} \notin A, \cdots, Y_1 \notin A | \boldsymbol{\xi}_0) = \boldsymbol{\xi} \boldsymbol{N}^{n-1}\boldsymbol{e}_i'. \qquad (2.22)$$

Proof. Utilizing the same arguments as in the proof of Theorem 2.2 and replacing $\boldsymbol{1}'$ by $\boldsymbol{e}_i'$, Eq. (2.22) follows directly from Eqs. (2.20) and (2.21).

$\square$

Theorem 2.4 *For any absorbing state $j \in A$, the probability of the system first entering the absorbing state j at the n-th trial is*

$$P(Y_n = j, Y_{n-1} \notin A, \cdots, Y_1 \notin A | \boldsymbol{\xi}_0) = \boldsymbol{\xi} \boldsymbol{N}^{n-1}\boldsymbol{C}_j', \qquad (2.23)$$

where $\boldsymbol{C}_j'$ is the j-th column of matrix $\boldsymbol{C}$.

Proof.   For any $j \in A$, it follows from the definition of the Markov chain and Theorem 2.3 that

$$P(Y_n = j, Y_{n-1} \notin A, \cdots, Y_1 \notin A|\boldsymbol{\xi}_0)$$

$$= \sum_{i \in \Omega - A} P(Y_{n-1} = i, Y_{n-2} \notin A, \cdots, Y_1 \notin A|\boldsymbol{\xi}_0)$$

$$\times P(Y_n = j|Y_{n-1} = i)$$

$$= \sum_{i \in \Omega - A} \boldsymbol{\xi} \boldsymbol{N}^{n-1} \boldsymbol{e}'_i p_{ij}$$

$$= \boldsymbol{\xi} \boldsymbol{N}^{n-1} \sum_{i \in \Omega - A} \boldsymbol{e}'_i p_{ij} = \boldsymbol{\xi} \boldsymbol{N}^{n-1} \boldsymbol{C}'_j.$$

$\square$

To illustrate Theorems 2.2 to 2.4 and their relationships, we provide the following simple example.

Example 2.5   Let $\{Y_t\}$ be a homogeneous Markov chain defined on the state space $\Omega = \{1, 2, 3, 4\}$ with initial distribution $\boldsymbol{\xi}_0 = (\boldsymbol{\xi} : \boldsymbol{0}) = (1, 0 : 0, 0)$ and transition probability matrix

$$\boldsymbol{M} = \begin{matrix} 1 \\ 2 \\ 3 \\ 4 \end{matrix} \begin{pmatrix} 1/2 & 1/4 & 1/4 & 0 \\ 1/3 & 1/3 & 0 & 1/3 \\ \hline 0 & 0 & 1 & 0 \\ 0 & 0 & 0 & 1 \end{pmatrix},$$

where $A = \{3, 4\}$ is the set of absorbing states. For $n = 3$, the first-entry probabilities of the system entering the absorbing states 3 and 4 are, respectively:

$$P(Y_3 = 3, Y_2 \notin A, Y_1 \notin A|\boldsymbol{\xi}_0) = (1, 0) \begin{pmatrix} 1/2 & 1/4 \\ 1/3 & 1/3 \end{pmatrix}^2 \begin{pmatrix} 1/4 \\ 0 \end{pmatrix} = \frac{1}{12},$$

and

$$P(Y_3 = 4, Y_2 \notin A, Y_1 \notin A|\boldsymbol{\xi}_0) = (1, 0) \begin{pmatrix} 1/2 & 1/4 \\ 1/3 & 1/3 \end{pmatrix}^2 \begin{pmatrix} 0 \\ 1/3 \end{pmatrix} = \frac{5}{72}.$$

Further, by Theorem 2.2, the probability that the system first enters the subset A at the third trial is

$$P(Y_3 \in A, Y_2 \notin A, Y_1 \notin A|\boldsymbol{\xi}_0) = \boldsymbol{\xi} \boldsymbol{N}^{n-1}(\boldsymbol{I} - \boldsymbol{N})\mathbf{1}'$$

$$= (1, 0) \begin{pmatrix} 1/2 & 1/4 \\ 1/3 & 1/3 \end{pmatrix}^2 \left[\begin{pmatrix} 1 & 0 \\ 0 & 1 \end{pmatrix} - \begin{pmatrix} 1/2 & 1/4 \\ 1/3 & 1/3 \end{pmatrix} \right] \begin{pmatrix} 1 \\ 1 \end{pmatrix} = \frac{11}{72}.$$

As a check on the above results, note that

$$P(Y_3 \in A, Y_2 \notin A, Y_1 \notin A|\boldsymbol{\xi}_0) = P(Y_3 = 3, Y_2 \notin A, Y_1 \notin A|\boldsymbol{\xi}_0)$$
$$+ P(Y_3 = 4, Y_2 \notin A, Y_1 \notin A|\boldsymbol{\xi}_0) = \frac{11}{72}.$$

$\diamond$

More generally, since for every $i \in \Omega - A$,

$$\sum_{j \in A} p_{ij} = 1 - \sum_{l \in \Omega - A} p_{il},$$

it follows that $\sum_{j \in A} \boldsymbol{C}_j' = (\boldsymbol{I} - \boldsymbol{N})\boldsymbol{1}'$, and hence

$$\sum_{j \in A} \boldsymbol{\xi} \boldsymbol{N}^{n-1} \boldsymbol{C}_j' = \boldsymbol{\xi} \boldsymbol{N}^{n-1} (\boldsymbol{I} - \boldsymbol{N}) \boldsymbol{1}'. \qquad (2.24)$$

2.7 First-Entry Probability

Using the ideas of Section 2.6, we can find the first-entry probability for any subset $B \subset \Omega$. Given the subset B, the transition probability matrix $\boldsymbol{M}$ of a homogeneous Markov chain $\{Y_t\}$ can again always be arranged in the following form:

$$\boldsymbol{M} = \begin{matrix} \Omega - B \\ B \end{matrix} \begin{pmatrix} \boldsymbol{N} & \boldsymbol{B} \\ \hline \boldsymbol{J} & \boldsymbol{Q} \end{pmatrix}. \qquad (2.25)$$

Theorem 2.5 *Let $\{Y_t\}$ be a homogeneous Markov chain with transition probability matrix $\boldsymbol{M}$, in the form of Eq. (2.25), and with initial distribution $\boldsymbol{\xi}_0 = (\boldsymbol{\xi} : \boldsymbol{0})$. Then, for a subset B of size k contained in the state space Ω of size m, the following relations hold true:*

(i) For all $j \in B$,

$$P(Y_n = j, Y_{n-1} \notin B, \cdots, Y_1 \notin B|\boldsymbol{\xi}_0) = \boldsymbol{\xi} \boldsymbol{N}^{n-1} \boldsymbol{B}_j', \qquad (2.26)$$

where $\boldsymbol{B}_j'$ is the j-th column of matrix $\boldsymbol{B}_{(m-k) \times k}$, and
(ii)

$$P(Y_n \in B, Y_{n-1} \notin B, \cdots, Y_1 \notin B|\boldsymbol{\xi}_0)$$
$$= \boldsymbol{\xi} \boldsymbol{N}^{n-1} (\boldsymbol{I} - \boldsymbol{N}) \boldsymbol{1}' = \sum_{j \in B} \boldsymbol{\xi} \boldsymbol{N}^{n-1} \boldsymbol{B}_j'. \qquad (2.27)$$

Proof.   Define a new Markov chain $\{Z_t\}$ on the state space Ω, where $\{Z_t\}$ is the same as $\{Y_t\}$ for all states $i \in \Omega - B$ and where all states $j \in B$ are taken as absorbing states. Then the transition probability matrix $\boldsymbol{M}^\star$ for the Markov chain $\{Z_t\}$ has the form

$$\boldsymbol{M}^\star = (p_{ij}^\star) = \left(\begin{array}{c|c} \boldsymbol{N} & \boldsymbol{B} \\ \hline \boldsymbol{O} & \boldsymbol{I} \end{array} \right).$$

Since, for each n,

$$P(Y_n = j, Y_{n-1} \notin B, \cdots, Y_1 \notin B | \boldsymbol{\xi}_0)$$
$$= P(Z_n = j, Z_{n-1} \notin B, \cdots, Z_1 \notin B | \boldsymbol{\xi}_0),$$

Result (i) follows from Theorem 2.4. Similarly, Result (ii) follows immediately from (i) and the fact that $(\boldsymbol{I} - \boldsymbol{N})\boldsymbol{1}' = \sum_{j \in B} \boldsymbol{B}_j'$. $\qquad \square$

The above proof is guided by the fact that all the states in B are absorbing states with respect to the new Markov chain $\{Z_t\}$, and hence, for example, for every $i \in \Omega - B$, the probability $P(Z_{n-1} = i)$ can be partitioned into two parts as follows:

$$P(Z_{n-1} = i | \boldsymbol{\xi}_0) = P(Z_{n-1} = i, Z_{n-2} \notin B, \cdots, Z_1 \notin B | \boldsymbol{\xi}_0)$$
$$+ P(Z_{n-1} = i \text{ and at least one of } Z_{n-2}, \cdots, Z_1 \text{ is in } B | \boldsymbol{\xi}_0),$$

where the second part is always zero (since $i \in \Omega - B$ and $p_{ji}^\star \equiv 0$ for all $j \in B$). Note that, in Theorem 2.5, we assumed the initial distribution $\boldsymbol{\xi}_0 = (\boldsymbol{\xi} : \boldsymbol{0})$, which is equivalent to saying $P(Y_0 \in \Omega - B) \equiv 1$. Consequently, the probability $P(Y_n \in B, Y_{n-1} \notin B, \cdots, Y_1 \notin B | \boldsymbol{\xi}_0)$ is referred to as the *first-entry* probability.

Example 2.6   Let $\{Y_t\}$ be a homogeneous Markov chain defined on the state space $\Omega = \{1, 2, 3, 4, 5\}$ with transition probability matrix

$$\boldsymbol{M} = \begin{array}{c} 1 \\ 2 \\ 3 \\ 4 \\ 5 \end{array} \left(\begin{array}{ccccc} 1/2 & 1/4 & 1/4 & 0 & 0 \\ 1/4 & 1/2 & 0 & 1/4 & 0 \\ 1/4 & 1/4 & 0 & 0 & 1/2 \\ 0 & 0 & 0 & 1 & 0 \\ 0 & 0 & 0 & 0 & 1 \end{array} \right).$$

Suppose $B = \{3, 4\}$, then the transition probability matrix of $\{Y_t\}$ can be re-arranged as

$$
\boldsymbol{M} = \begin{array}{c} 1 \\ 2 \\ 5 \\ 3 \\ 4 \end{array} \left(\begin{array}{ccc:cc} 1/2 & 1/4 & 0 & 1/4 & 0 \\ 1/4 & 1/2 & 0 & 0 & 1/4 \\ 0 & 0 & 1 & 0 & 0 \\ \hdashline 1/4 & 1/4 & 1/2 & 0 & 0 \\ 0 & 0 & 0 & 0 & 1 \end{array} \right).
$$

Given an initial distribution of Y_0, say $P(Y_0 = 1) = 1$, the first-entry probability for $Y_n = 3$ is

$$
P(Y_n = 3, Y_{n-1} \notin B, \cdots, Y_1 \notin B | Y_0 = 1)
$$

$$
= (1, 0, 0) \begin{pmatrix} 1/2 & 1/4 & 0 \\ 1/4 & 1/2 & 0 \\ 0 & 0 & 1 \end{pmatrix}^{n-1} \begin{pmatrix} 1/4 \\ 0 \\ 0 \end{pmatrix}.
$$

For $n = 3$, this yields $P(Y_3 = 3, Y_2 \notin B, Y_1 \notin B | Y_0 = 1) = 5/64$. Note that since state "5" is an absorbing state, the computations can be reduced further to

$$
P(Y_n = 3, Y_{n-1} \notin B, \cdots, Y_1 \notin B | Y_0 = 1) = (1, 0) \begin{pmatrix} 1/2 & 1/4 \\ 1/4 & 1/2 \end{pmatrix}^{n-1} \begin{pmatrix} 1/4 \\ 0 \end{pmatrix}.
$$

$\diamond$

Let A contain all of the absorbing states of $\{Y_t\}$, and let $B^* = A \cup B$. The transition probability matrix $\boldsymbol{M}^*$, corresponding to the associated Markov chain $\{Z_t\}$ in which all the states in B^* are taken as absorbing states, can then be re-arranged as

$$
\boldsymbol{M}^* = \begin{array}{c} \Omega - B^* \\ B^* \end{array} \left(\begin{array}{c:c} \boldsymbol{N}^* & \boldsymbol{B}^* \\ \hdashline \boldsymbol{O} & \boldsymbol{I} \end{array} \right),
$$

where $\boldsymbol{N}^*$ is the essential submatrix of $\boldsymbol{M}^*$, and $\boldsymbol{B}^*$ is the matrix formed from $\boldsymbol{M}^*$ by deletion of all columns associated with states in $\Omega - B^*$ and of all rows associated with states in B^*. The following corollary then holds.

Corollary 2.1   *For every $j \in B$,*

$$
P(Y_n = j, Y_{n-1} \notin B, \cdots, Y_1 \notin B | (\boldsymbol{\xi}^* : \boldsymbol{0})) = \boldsymbol{\xi}^* (\boldsymbol{N}^*)^{n-1} \boldsymbol{B}_j^{*\prime}, \qquad (2.28)
$$

where $\boldsymbol{B}_j^{*\prime}$ is the j-th column of the matrix $\boldsymbol{B}^*$.

The proof of the above corollary is straightforward, and left to the reader. Note that the size of the matrix N^* is smaller than or equal to the size of N.

Chapter 3

Runs and Patterns in a Sequence of Two-State Trials

3.1 Introduction

Although runs and patterns in a sequence of Bernoulli trials are special cases of those defined on multi-state trials, they merit a separate chapter due to their long history, the large quantity of associated results, and their broad application to numerous fields. The focus of this chapter will be on deriving the distributions for the most common and useful runs statistics in Bernoulli trials via the finite Markov chain imbedding technique, and also on extending these results to sequences of Markov-dependent two-state trials.

Techniques for obtaining the recursive equations and probability generating functions of runs statistics through the finite Markov chain imbedding approach are also introduced. These tools can be very useful for studying certain characteristics of the distributions of runs, such as the mean, variance, and higher moments.

The following runs statistics, defined traditionally on a sequence of n Bernoulli trials, are treated in this chapter:

(i) $N_{n,k}$, the number of non-overlapping consecutive k successes;

(ii) $G_{n,k}$, the number of success runs of size greater than or equal to k;

(iii) $M_{n,k}$, the number of overlapping consecutive k successes;

(iv) $E_{n,k}$, the number of runs of exactly k successes;

(v) $L_n(S)$, the size of the longest success run; and

(vi) $S_{n,k}$, the total number of successes in success runs of length greater than or equal to k.

The waiting-time distribution of a success run is also treated, and some numerical results for the distributions of the above runs statistics are included.

3.2   Number of Non-Overlapping Consecutive k Successes

The number of runs of non-overlapping consecutive k successes, $N_{n,k}$, in a sequence of n Bernoulli trials is probably the most important runs statistic, not only for its broad application to various areas but also for its connection with other runs statistics; in distribution theory, the distribution of $N_{n,k}$ is referred to as the Binomial distribution of order k. A formula for the exact distribution of $N_{n,k}$ in a sequence of n Bernoulli trials was given independently by Philippou and Makri (1986) and by Hirano (1986) as

$$P(N_{n,k} = x) = \sum_{m=0}^{k-1} \sum_{\substack{x_1+2x_2+\cdots+ \\ kx_k=n-m-kx}} \binom{x_1+\cdots+x_k+x}{x_1,\cdots,x_k,x} p^n (\frac{q}{p})^{x_1+\cdots+x_k},$$

(3.1)

where $x = 0, 1, \cdots, [n/k]$ ($[n/k]$ is the integer part of n/k), and the success and failure probabilities are denoted by p and $q = 1 - p$, respectively. Godbole (1990) gave an alternative formula for the *pdf* of $N_{n,k}$ with $k > 1$:

$$P(N_{n,k} = x) = \sum_{[(n-kx)/k]\leq y\leq n-kx} q^y p^{n-y} \binom{y+x}{x}$$
$$\times \sum_{0\leq j\leq[(n-kx-y)/k]} (-1)^j \binom{y+1}{j} \binom{n-kx-jk}{y},$$

(3.2)

for $x = 0, 1, \cdots, [n/k]$. Formula (3.2) has an advantage over (3.1) in that it is easier to evaluate by computer for large n. Hirano and Aki (1987, 1993) studied some properties of this distribution, and extended the results to the case of two-state Markov-dependent trials.

To begin our study of $N_{n,k}$ using the method of finite Markov chain imbedding, let us consider the state space

$$\Omega = \{(x,i) : x = 0, 1, \cdots, l_n, \text{ and } i = 0, 1, \cdots, k-1\},$$

where $l_n = [n/k]$ is the maximum number of non-overlapping success runs of length k that can occur within the n trials. We define a finite homogeneous

Markov chain $\{Y_t : t = 0, 1, \cdots, n\}$ on Ω as follows:

$$Y_t = (N_{t,k}, E_t), \text{ for } 1 \le t \le n, \tag{3.3}$$

where $N_{t,k}$ is the number of non-overlapping consecutive k successes that occurred over the first t trials $X_1, \cdots, X_t$. The "ending block" E_t equals m modulo k, where m represents the number of trailing successes (possibly zero) that exist in the sequence after the first t trials:

$$FFSSF \underbrace{SS \cdots S}_{m}.$$

Note that $E_t = 0$ either if m is a positive multiple of k or if the t-th outcome is F. This ending-block variable keeps track of the number of successes in a possible partial run associated with the t-th trial. For example, given $n = 10$ Bernoulli trials with outcomes $\{FSFFSSSFSS\}$ and a chosen success run length of $k = 2$, the realization of the imbedded Markov chain $\{Y_t : t = 1, 2, \cdots, 10\}$ with respect to these ten outcomes is: $\{Y_1 = (0,0), Y_2 = (0,1), Y_3 = (0,0), Y_4 = (0,0), Y_5 = (0,1), Y_6 = (1,0), Y_7 = (1,1), Y_8 = (1,0), Y_9 = (1,1), Y_{10} = (2,0)\}$. Note that for a given sequence of outcomes $\{FS \cdots SF\}$, the realization of $\{Y_t\}$ is always unique.

Define the subsets

$$C_x = \{(x,i) : i = 0, 1, \cdots, k-1\}, \ 0 \le x \le l_n. \tag{3.4}$$

The collection of subsets $\{C_x : x = 0, 1, \cdots, l_n\}$ forms a partition of the state space Ω. Since $\{X_t\}$ is, for the moment, a sequence of Bernoulli trials, it follows from the above definitions that Y_t has a transition probability matrix $\boldsymbol{M}_t = (p_{(x,i)(y,j)})$ for all $t = 1, 2, \cdots, n$, with the transition probabilities $p_{(x,i)(y,j)}$ given by the following equation: for $1 \le t \le n$ and $0 \le x \le l_n$,

$$
\begin{aligned}
p_{(x,i)(y,j)} \ &= \ P(Y_t = (y,j)|Y_{t-1} = (x,i)) \\
&= \begin{cases}
q & \text{if } y = x \text{ and } j = 0, \text{ for } i = 0, 1, \cdots, k-1 \\
p & \text{if } y = x \text{ and } j = i+1, \text{ for } i = 0, 1, \cdots, k-2 \\
p & \text{if } y = x+1 \text{ and } j = 0, \text{ for } i = k-1, \text{ and} \\
& \quad x = 0, 1, \cdots, l_n - 1 \\
1 & \text{if } y = x = l_n \text{ and } j = i = k-1 \\
0 & \text{otherwise.}
\end{cases}
\end{aligned}
\tag{3.5}
$$

As an illustration, the transition probability matrix $\boldsymbol{M}_t$ of the imbedded Markov chain $\{Y_t\}$ associated with the random variable $N_{5,2}$ is given by

$$
\boldsymbol{M}_t(N_{5,2}) =
\begin{array}{c}
(0,0) \\ (0,1) \\ (1,0) \\ (1,1) \\ (2,0) \\ (2,1)
\end{array}
\left(
\begin{array}{cc:cc:cc}
q & p & 0 & 0 & 0 & 0 \\
q & 0 & p & 0 & 0 & 0 \\ \hdashline
0 & 0 & q & p & 0 & 0 \\
0 & 0 & q & 0 & p & 0 \\ \hdashline
0 & 0 & 0 & 0 & q & p \\
0 & 0 & 0 & 0 & 0 & 1
\end{array}
\right)_{6\times 6} , \qquad (3.6)
$$

for $1 \le t \le 5$.

For the case where $\{X_t\}$ is a sequence of independent but non-identically distributed two-state trials with probabilities $P(X_t = S) = p_t$ and $P(X_t = F) = q_t$, for $t = 1, \cdots, n$, the transition matrices $\boldsymbol{M}_t$ for the imbedded Markov chain $\{Y_t\}$ remain unchanged except that the probability p is replaced by p_t and q is replaced by q_t. In general for independent but non-identically distributed two-state trials, the transition probability matrices can be written as

$$
\boldsymbol{M}_t(N_{n,k}) =
\begin{pmatrix}
\boldsymbol{A}_t & \boldsymbol{B}_t & & & \boldsymbol{0} \\
& \ddots & \ddots & & \\
\boldsymbol{0} & & & \boldsymbol{A}_t & \boldsymbol{B}_t \\
& & & & \boldsymbol{A}_t^*
\end{pmatrix}_{d\times d} , \qquad (3.7)
$$

for $t = 1, 2, \cdots, n$, where

$$
\boldsymbol{A}_t =
\begin{pmatrix}
q_t & p_t & 0 & \cdots & 0 \\
q_t & 0 & p_t & & \\
\vdots & & \ddots & \ddots & \\
\vdots & & & \ddots & p_t \\
q_t & 0 & 0 & \cdots & 0
\end{pmatrix}_{k\times k} ,
$$

$\boldsymbol{B}_t$ is a $k \times k$ matrix having p_t at the entry $(k, 1)$ and zero elsewhere, $\boldsymbol{A}_t^*$ is the same as $\boldsymbol{A}_t$ except for having its last row replaced with $(0, 0, \cdots, 0, 1)$, and the dimension of $\boldsymbol{M}_t(N_{n,k})$ is given by $d = k(l_n + 1)$. Hence, by virtue of Theorem 2.1, we may state that

$$
P(N_{n,k} = x) = \boldsymbol{\xi}_0 \Big(\prod_{t=1}^{n} \boldsymbol{M}_t(N_{n,k}) \Big) \boldsymbol{U}'(C_x), \quad x = 0, 1, \cdots, l_n, \qquad (3.8)
$$

where $\boldsymbol{\xi}_0 = (1, 0, \cdots, 0)_{1 \times d}$, and $\boldsymbol{U}'(C_x)$ is the transpose of the vector $\boldsymbol{U}(C_x) = (0, \cdots, 0, 1, \cdots, 1, 0, \cdots, 0)$ with ones at the locations associated with the states in C_x. Equation (3.8) yields the exact distribution of $N_{n,k}$ for both identically as well as non-identically distributed independent two-state trials. In view of the transition probability matrix in Eq. (3.7), $N_{n,k}$ is finite Markov chain imbeddable of binomial type in the sense of Definition 2.7 (Koutras and Alexandrou 1995).

If $\{X_t\}$ is a non-homogeneous Markov chain with transition probability matrices

$$\begin{pmatrix} p_{FF}(t) & p_{FS}(t) \\ p_{SF}(t) & p_{SS}(t) \end{pmatrix},$$

a minor modification to the imbedding procedure is needed to obtain the distribution of $N_{n,k}$. Since the outcome of X_{t+1}, and hence also of Y_{t+1}, now depends on X_t, each state of the Markov chain Y_t must imply a certain outcome of X_t. This is already the case in our above definition of Y_t, save for the states with an ending block of $E_t = 0$, which can arise for either outcome of X_t. To resolve this ambiguity, we define an additional ending-block state, $E_t = \gamma$, to correspond to the case where the run of trailing successes is a non-zero multiple of k successes, and reserve the state $E_t = 0$ for the case where the t-th outcome is a failure. The imbedded Markov chain is then defined as follows:

$Y_t = (x, \gamma)$
$\Rightarrow$ there are x runs of k consecutive successes in the first t trials with $m > 0$ trailing successes such that $m \bmod k = 0$,

$Y_t = (x, 0)$
$\Rightarrow$ there are x runs of k consecutive successes in the first t trials with $m = 0$ trailing successes $(X_t = F)$,

and $Y_t = (x, i)$, for $i = 1, \cdots, k - 1$, is defined as given in Eq. (3.3). The difference between the states (x, γ) and $(x, 0)$ can be seen from the following example: for a success run length of $k = 2$, $Y_8(SFFFSSSS) = (2, \gamma)$ and $Y_8(SFSSFSSF) = (2, 0)$. Note that the ending block E_t now contains not only the required information on subpatterns but also implies the outcome of X_t, enabling the assignment of transition probabilities for the imbedded Markov chain.

The transition probability matrices corresponding to these definitions may then be readily constructed. The imbedded Markov chain associated with the random variable $N_{5,2}$, as considered in Eq. (3.6) for Bernoulli trials, has the following transition matrices $\boldsymbol{M}_t^*$ under non-homogeneous Markov-dependent trials: for $t = 1, \cdots, n$,

$$
\boldsymbol{M}_t^*(N_{5,2}) =
\begin{array}{c}
\begin{array}{cccccccc}
(0,0) & (0,1) & (1,\gamma) & (1,0) & (1,1) & (2,\gamma) & (2,0) & (2,1)
\end{array} \\
\left(
\begin{array}{cc:ccc:ccc}
p_{FF}(t) & p_{FS}(t) & 0 & 0 & 0 & 0 & 0 & 0 \\
p_{SF}(t) & 0 & p_{SS}(t) & 0 & 0 & 0 & 0 & 0 \\ \hdashline
0 & 0 & 0 & p_{SF}(t) & p_{SS}(t) & 0 & 0 & 0 \\
0 & 0 & 0 & p_{FF}(t) & p_{FS}(t) & 0 & 0 & 0 \\
0 & 0 & 0 & p_{SF}(t) & 0 & p_{SS}(t) & 0 & 0 \\ \hdashline
0 & 0 & 0 & 0 & 0 & 0 & p_{SF}(t) & p_{SS}(t) \\
0 & 0 & 0 & 0 & 0 & 0 & p_{FF}(t) & p_{FS}(t) \\
0 & 0 & 0 & 0 & 0 & 0 & 0 & 1
\end{array}
\right)
\end{array}.
$$

Note the similar banded structure of $\boldsymbol{M}_t^*(N_{5,2})$ in comparison to $\boldsymbol{M}_t(N_{5,2})$ of Eq. (3.6) for Bernoulli trials. As it is straightforward to derive the general form of $\boldsymbol{M}_t^*(N_{n,k})$ analogous to Eq. (3.7), we leave this to the interested reader.

When the sequence $\{X_t\}$ is *i.i.d.*, the initial distribution $\boldsymbol{\xi}_0$ can be defined by $P(Y_0 = (0,0)) = 1$, yielding, for $k > 1$, the transition probabilities $P(Y_1 = (0,1)|Y_0 = (0,0)) = p$ and $P(Y_1 = (0,0)|Y_0 = (0,0)) = q = (1 - p)$. However, when $\{X_t\}$ is a sequence of Markov-dependent random variables, one must be careful about assuming $P(Y_0 = (0,0)) = 1$, which would imply that the transition probabilities between Y_0 and Y_1 are given by $P(Y_1 = (0,1)|Y_0 = (0,0)) = p_{FS}(1)$ and $P(Y_1 = (0,0)|Y_0 = (0,0)) = p_{FF}(1)$, independent of p_{SF} and p_{SS}. In order to avoid this type of bias, it is useful to create a dummy state $\emptyset$ as the initial state for Y_0. We then define $P(Y_0 = \emptyset) = 1$, and the transition probabilities $P(Y_1 = (0,1)|Y_0 = \emptyset) = p_S$ and $P(Y_1 = (0,0)|Y_0 = \emptyset) = p_F$. It follows that, for $N_{5,2}$, the corresponding imbedded Markov chain $\{Y_t\}$ is defined on the expanded state space $\Omega = \{\emptyset, (0,0), (0,1), (1,0), \cdots\}$ with transition probability matrices of the following form:

$$
\begin{array}{c}
\begin{array}{cccccc}
\emptyset & (0,0) & (0,1) & (1,0) & (1,1) & \cdots
\end{array} \\
\left(
\begin{array}{c:ccccc}
0 & p_F & p_S & 0 & 0 & \cdots \\ \hdashline
0 & & & & & \\
\vdots & & & \boldsymbol{M}_t^*(N_{5,2}) & & \\
0 & & & & &
\end{array}
\right)
\end{array}.
$$

Note that the finite Markov chain imbedding procedure used to obtain the exact distribution of $N_{n,k}$ remains the same, except for minor differences in the transition probability matrices, regardless of whether the sequence of trials $\{X_t\}$ is *i.i.d.*, independent but non-identically distributed, or Markov-dependent.

3.3 Number of Success Runs of Length Greater Than or Equal to k

For a sequence of two-state trials, the random variable $G_{n,k}$ is defined as the number of success runs of length greater than or equal to k. Let's consider a finite Markov chain $\{Y_t : t = 0, 1, \cdots, n\}$ defined on the state space

$$\Omega = \{(x,i) : x = 0, 1 \cdots, l_n, \text{and } i = \gamma, 0, 1, \cdots, k-1\} - \{(0,\gamma)\},$$

where $l_n = [(n+1)/(k+1)]$. For a sequence of outcomes of the first t trials with m trailing successes, say $FS \cdots F \underbrace{SS \cdots S}_{m}$, we define the Markov chain

$$Y_t = (G_{t,k}, E_t), \quad 1 \le t \le n, \tag{3.9}$$

where $G_{t,k}$ is the number of success runs of length greater than or equal to k in the sequence $\{X_t\}$, and E_t is the ending block variable with $E_t = m$ if $m = 0, 1, \cdots, k-1$, and $E_t = \gamma$ if $m \ge k$. To illustrate this definition, consider a minimum run-length of $k = 2$ and the following twelve outcomes of two-state trials: $FSFFSSFSSSFS$, for which $G_{12,2} = 2$. It follows from Eq. (3.9) that the realization of the Markov chain $\{Y_t : t = 1, \cdots, 12\}$ is $\{Y_1 = (0,0), Y_2 = (0,1), Y_3 = (0,0), Y_4 = (0,0), Y_5 = (0,1), Y_6 = (1,\gamma), Y_7 = (1,0), Y_8 = (1,1), Y_9 = (2,\gamma), Y_{10} = (2,\gamma), Y_{11} = (2,0),$ and $Y_{12} = (2,1)\}$. Note that the ending block $E_t = \gamma$ can occur only when there are at least k trailing successes, in which case $G_{n,k} \ge 1$; for this reason, the state $(0,\gamma)$ was excluded in the above definition of the state space Ω.

From the definition of the imbedded Markov chain given by Eq. (3.9), the one-step transition probabilities in $M_t(G_{n,k})$ for independent but non-identically distributed trials are specified by the following equation: for

$t = 1, \cdots, n,$

$$p_{(x,i)(y,j)}(t) = \begin{cases} q_t & \text{if } y = x \text{ and } j = 0, \text{ for } x = 0, 1, \cdots, l_n, \text{ and } i = \gamma, 0, 1, \\ & \cdots, k-1 \\ p_t & \text{if } y = x \text{ and } j = i = \gamma, \text{ for } x = 0, 1, \cdots, l_n \\ p_t & \text{if } y = x \text{ and } j = i+1, \text{ for } x = 0, 1, \cdots, l_n, \text{ and } i = \gamma, 0, \\ & 1, \cdots, k-2 \\ p_t & \text{if } y = x+1, i = k-1 \text{ and } j = \gamma, \text{ for } x = 0, 1, \cdots, \\ & l_n - 1 \\ 1 & \text{if } y = x = l_n \text{ and } j = x = k-1 \\ 0 & \text{otherwise.} \end{cases}$$

$$(3.10)$$

For the special case of $n = 5$ and $k = 2$, the transition probability matrices $\boldsymbol{M}_t(G_{n,k})$ are given by

$$\boldsymbol{M}_t(G_{5,2}) = \begin{array}{c} (0,0) \\ (0,1) \\ (1,\gamma) \\ (1,0) \\ (1,1) \\ (2,\gamma) \\ (2,0) \\ (2,1) \end{array} \left(\begin{array}{cc|cc|cc|cc} q_t & p_t & 0 & 0 & 0 & 0 & 0 & 0 \\ q_t & 0 & p_t & 0 & 0 & 0 & 0 & 0 \\ 0 & 0 & p_t & q_t & 0 & 0 & 0 & 0 \\ 0 & 0 & 0 & q_t & p_t & 0 & 0 & 0 \\ 0 & 0 & 0 & q_t & 0 & p_t & 0 & 0 \\ 0 & 0 & 0 & 0 & 0 & p_t & q_t & 0 \\ 0 & 0 & 0 & 0 & 0 & 0 & q_t & p_t \\ 0 & 0 & 0 & 0 & 0 & 0 & 0 & 1 \end{array} \right),$$

for $t = 1, \cdots, 5$.

In general, $\boldsymbol{M}_t(G_{n,k})$ is a bi-diagonal block matrix of the form

$$\boldsymbol{M}_t(G_{n,k}) = \begin{pmatrix} \boldsymbol{A}_t & p_t \boldsymbol{e}'_k & & & & & \boldsymbol{O} \\ & p_t & q_t \boldsymbol{e}_1 & & \boldsymbol{O} & & \\ & & \boldsymbol{A}_t & p_t \boldsymbol{e}'_k & & & \\ & & & \ddots & \ddots & & \\ & & & & \ddots & \ddots & \\ & \boldsymbol{O} & & & & p_t & q_t \boldsymbol{e}_1 \\ \boldsymbol{O} & & & & & & \boldsymbol{A}_t^* \end{pmatrix}, \quad (3.11)$$

where $\boldsymbol{e}_1 = (1, 0, \cdots, 0)$ and $\boldsymbol{e}_k = (0, \cdots, 0, 1)$ are $1 \times k$ unit row vectors, and $\boldsymbol{A}_t$ is given by

$$
\boldsymbol{A}_t = \begin{pmatrix}
q_t & p_t & 0 & \cdots & 0 \\
\vdots & \ddots & \ddots & & \\
\vdots & & \ddots & \ddots & \\
\vdots & & & \ddots & p_t \\
q_t & 0 & 0 & \cdots & 0
\end{pmatrix}.
$$

The transition probability matrix $\boldsymbol{A}_t$, in the context of demography, is often referred to as the Leslie matrix, or more generally, as a renewal-type matrix (see Seneta 1981). The dimension of $\boldsymbol{M}_t(G_{n,k})$ is equal to $(l_n+1)(k+1)-1$. The matrix $\boldsymbol{A}_t^*$ in Eq. (3.11) is the same as $\boldsymbol{A}_t$ except for the last row, which is replaced by $(0,0,\cdots,0,1)$.

We define the partition $\{C_x : x = 0, 1, \cdots, l_n\}$ for Ω as

$$
\begin{aligned}
C_0 &= \{(0,i) : i = 0,1,\cdots,k-1\}, \\
C_x &= \{(x,i) : i = \gamma, 0, 1, \cdots, k-1\}, \text{ for } x = 1, \cdots, l_n,
\end{aligned}
$$

from which it follows that $P(G_{n,k} = x) = P(Y_n \in C_x)$ for all $x = 0, 1, \cdots, l_n$. The distribution function, moments, and probability generating function can now be easily computed through Eqs. (2.11), (2.12) and (2.13), respectively.

For the case of *i.i.d.* trials, all transition probabilities would be constant, and an extension to Markov-dependent trials may be carried out as described for the statistic $N_{n,k}$ in the previous section; in the remainder of this chapter on two-state trials, we focus primarily on the case of independent but non-identically distributed trials.

3.4   Number of Overlapping Consecutive k Successes

The random variable $M_{n,k}$ is defined as the number of *overlapping* consecutive k successes in a sequence of n independent two-state trials. The imbedded Markov chain $\{Y_t : t = 0, 1, \cdots, n\}$ associated with $M_{n,k}$ may be defined as

$$
Y_t = (M_{t,k}, E_t), \quad t = 1, 2, \cdots, n, \tag{3.12}
$$

on the state space

$$
\begin{aligned}
\Omega \;=\; & \{(x,i) : x = 0, 1, \cdots, l_n - 1, \text{and } i = \gamma, 0, 1, \cdots, k - 1\} \\
& \cup \{(l_n, \gamma)\} - \{(0, \gamma)\},
\end{aligned}
$$

where $l_n = n - k + 1$, $M_{t,k}$ is the number of overlapping consecutive k successes in the first t trials, and E_t is the ending block variable keeping track of the number of trailing successes m:

$$
E_t = \left\{ \begin{array}{ll} \gamma & \text{if } m \geq k \\ m & \text{if } m = 0, 1, \cdots, k - 1. \end{array} \right. \tag{3.13}
$$

With overlap-counting, it is easy to verify that the probabilities for the transition probability matrices $\boldsymbol{M}_t = (p_{(x,i)(y,j)}(t))$ can be obtained from the following equation:

$$
p_{(x,i)(y,j)}(t) = \left\{ \begin{array}{ll} q_t & \text{if } y = x \text{ and } j = 0, \text{ for } x = 0, 1 \cdots, l_n, \text{ and } i = \gamma, 0, 1, \\ & \cdots, k - 1 \\ p_t & \text{if } y = x \text{ and } j = i + 1, \text{ for } x = 0, 1 \cdots, l_n, \text{ and } i = 0, 1, \\ & \cdots, k - 2 \\ p_t & \text{if } y = x + 1, \ j = \gamma \text{ and } i = k - 1, \text{ for } x = 0, 1, \cdots, \\ & l_n - 1 \\ p_t & \text{if } y = x + 1 \text{ and } j = i = \gamma, \text{ for } x = 0, 1 \cdots, l_n - 1 \\ 1 & \text{if } y = x = l_n \text{ and } j = i = \gamma \\ 0 & \text{otherwise.} \end{array} \right.
$$

$$\tag{3.14}$$

The corresponding partition of the state space Ω can be specified as follows:

$$
\begin{aligned}
C_0 \;&=\; \{(0, i) : i = 0, 1, \cdots, k - 1\}, \\
C_x \;&=\; \{(x, i) : i = \gamma, 0, 1, \cdots, k - 1\}, \ x = 1, \cdots, l_n - 1, \\
C_{l_n} \;&=\; \{(l_n, \gamma)\}.
\end{aligned}
$$

For $n = 4$ and $k = 2$, for example, the transition probability matrices

$M_t(M_{4,2})$, $t = 1, 2, 3, 4$, are

$$
M_t(M_{4,2}) =
\begin{array}{c}
(0,0) \\
(0,1) \\
(1,\gamma) \\
(1,0) \\
(1,1) \\
(2,\gamma) \\
(2,0) \\
(2,1) \\
(3,\gamma)
\end{array}
\left(
\begin{array}{ccc|ccc|ccc}
q_t & p_t & 0 & 0 & 0 & 0 & 0 & 0 & 0 \\
q_t & 0 & p_t & 0 & 0 & 0 & 0 & 0 & 0 \\
0 & 0 & 0 & q_t & 0 & p_t & 0 & 0 & 0 \\
0 & 0 & 0 & q_t & p_t & 0 & 0 & 0 & 0 \\
0 & 0 & 0 & q_t & 0 & p_t & 0 & 0 & 0 \\
0 & 0 & 0 & 0 & 0 & 0 & q_t & 0 & p_t \\
0 & 0 & 0 & 0 & 0 & 0 & q_t & p_t & 0 \\
0 & 0 & 0 & 0 & 0 & 0 & q_t & 0 & p_t \\
0 & 0 & 0 & 0 & 0 & 0 & 0 & 0 & 1
\end{array}
\right) . \tag{3.15}
$$

For general n and k, the transition probability matrices continue to have a banded form similar to $M_t(M_{4,2})$ in Eq. (3.15), and are of dimension $l_n(k+1)$. The distribution and moments for the random variable $M_{n,k}$ can again be computed through Eqs. (2.11) and (2.12), respectively.

3.5 Number of Runs of Exactly k Successes

The imbedded Markov chain $\{Y_t\}$ associated with the random variable $E_{n,k}$, the number of success runs of size exactly k in n independent two-state trials, is defined by

$$
Y_t = (E_{t,k}, E_t) \tag{3.16}
$$

on the state space

$$
\Omega = \{(x, i) : x = 0, 1, \cdots, l_n, \text{ and } i = \beta, \gamma, 0, 1, \cdots, k-1\} - \{(0,\gamma)\},
$$

where $l_n = [(n+1)/(k+1)]$, $E_{t,k}$ is the number of success runs of size exactly k in the first t trials, and the ending block E_t is defined according to the number of trailing successes m in the first t trials as follows:

$$
E_t = \begin{cases}
m & \text{if } m = 0, 1, \cdots, k-1 \\
\gamma & \text{if } m = k \\
\beta & \text{if } m > k.
\end{cases} \tag{3.17}
$$

The two ending-block states β and γ carry the following interpretation:

(i) Waiting state (x, γ), $x = 1, \cdots, l_n$:
  $Y_t = (x, \gamma)$ means that $m = k$ and that the x-th success run of size

k has occurred at the t-th trial, and

(ii) Overflow state (x, β), $x = 1, \cdots, l_n$:
    $Y_t = (x, \beta)$ means that $m > k$ and that exactly x success runs of size k have appeared prior to the last $m + 1$ outcomes $(F \underbrace{S \cdots S}_{m})$.

With these ending blocks in mind, we can easily construct the partition for the state space Ω: $C_0 = \{(0, i) : i = \beta, 0, 1, \cdots, k - 1\}$, and $C_x = \{(x, i) : i = \gamma, \beta, 0, 1, \cdots, k - 1\}$, for $x = 1, \cdots, l_n$.

The probabilities for the transition matrices $\boldsymbol{M}_t(E_{t,k})$ of the imbedded Markov chain $\{Y_t\}$ are specified by the following equation:

$$
P_{(x,i)(y,j)}(t) = \begin{cases}
q_t & \text{if } y = x \text{ and } j = 0, \text{ for } x = 0, 1, \cdots, l_n, \text{ and } i = \gamma, \beta, 0, 1, \\
& \quad \cdots, k - 1 \\
p_t & \text{if } y = x \text{ and } j = i + 1, \text{ for } x = 0, 1 \cdots, l_n, \text{ and } i = 0, 1, \\
& \quad \cdots, k - 2 \\
p_t & \text{if } y = x + 1, j = \gamma \text{ and } i = k - 1, \text{ for } x = 0, 1, \cdots, l_n - 1 \\
p_t & \text{if } y = x - 1, j = \beta \text{ and } i = \gamma, \text{ for } x = 1, \cdots, l_n \\
p_t & \text{if } y = x \text{ and } j = i = \beta, \text{ for } x = 0, 1 \cdots, l_n \\
1 & \text{if } y = x = l_n \text{ and } j = i = k - 1 \\
0 & \text{otherwise.}
\end{cases}
$$

$$(3.18)$$

As an illustration, consider the case $n = 5$ and $k = 2$, for which we have the transition probability matrices

$$
\boldsymbol{M}_t(E_{5,2}) = \begin{array}{c}
(0, \beta) \\
(0, 0) \\
(0, 1) \\
(1, \gamma) \\
(1, \beta) \\
(1, 0) \\
(1, 1) \\
(2, \gamma) \\
(2, \beta) \\
(2, 0) \\
(2, 1)
\end{array}
\left(
\begin{array}{cccccccccc}
q_t & p_t & 0 & 0 & 0 & 0 & 0 & 0 & 0 & 0 \\
0 & q_t & p_t & 0 & 0 & 0 & 0 & 0 & 0 & 0 \\
0 & q_t & 0 & p_t & 0 & 0 & 0 & 0 & 0 & 0 \\
p_t & 0 & 0 & 0 & 0 & q_t & 0 & 0 & 0 & 0 \\
0 & 0 & 0 & 0 & p_t & q_t & 0 & 0 & 0 & 0 \\
0 & 0 & 0 & 0 & 0 & q_t & p_t & 0 & 0 & 0 \\
0 & 0 & 0 & 0 & 0 & q_t & 0 & p_t & 0 & 0 \\
0 & 0 & 0 & 0 & p_t & 0 & 0 & 0 & q_t & 0 \\
0 & 0 & 0 & 0 & 0 & 0 & 0 & 0 & p_t & q_t & 0 \\
0 & 0 & 0 & 0 & 0 & 0 & 0 & 0 & q_t & p_t \\
0 & 0 & 0 & 0 & 0 & 0 & 0 & 0 & 0 & 1
\end{array}
\right).
$$

$$(3.19)$$

In general, the transition probability matrices of the Markov chain $\{Y_t\}$ associated with $E_{n,k}$ have the form given by Eq. (3.19) with dimension $(l_n + 1)(k + 1) + l_n$.

3.6 The Distribution of the Longest Success Run

Let $L_n(S)$ be the length of the longest success run in a sequence of two-state trials. For the case of n independent tosses of a fair coin, let $A_n(k)$ be the number of sequences of length n in which the longest run of successes (heads) is less than or equal to k. Since all the sequences are equally likely with probability $1/2^n$, the distribution of the longest run is

$$P(L_n(S) \leq k) = 2^{-n} A_n(k), \tag{3.20}$$

where $A_n(k)$ satisfies the recursive equation (Schilling 1990)

$$A_n(k) = \begin{cases} \sum_{j=0}^{k} A_{n-1-j}(k) & \text{if } n > k \\ 2^n & \text{if } n \leq k \\ 1 & \text{if } k = 0. \end{cases} \tag{3.21}$$

For biased coins $(p \neq 1/2)$, combinatorial analysis yields

$$P(L_n(S) \leq k) = \sum_{x=0}^{n} C_n^{(x)}(k) p^x q^{n-x}, \tag{3.22}$$

for $1 \leq k \leq n$, and $P(L_n(S) = 0) = q^n$, where $C_n^{(x)}(k)$ is the number of sequences of length n in which exactly x successes occur, but in which no more than k of these successes occur consecutively. $C_n^{(x)}(k)$ can be obtained through the recursive equation

$$C_n^{(x)}(k) = \begin{cases} \sum_{j=0}^{k} C_{n-1-j}^{(x-j)}(k) & \text{if } k < x < n \\ \binom{n}{x} & \text{if } x \leq k \leq n \\ 0 & \text{if } k < x = n. \end{cases} \tag{3.23}$$

More generally, suppose that the probabilities of success and failure could be different in each trial, equal to p_t and q_t, respectively, for $t = 1, \cdots, n$. The following theorem derives the distribution for $L_n(S)$:

Theorem 3.1   *For $0 \le k \le n$,*

$$P(L_n(S) \le k) = \boldsymbol{\xi}(\prod_{t=1}^{n} \boldsymbol{N}_t)\boldsymbol{1}'_{1 \times (k+1)}, \tag{3.24}$$

where $\boldsymbol{\xi} = (1, 0, \cdots, 0)$ is a $1 \times (k+1)$ unit row vector, and $\boldsymbol{N}_t$ is, as indicated below, the $(k+1) \times (k+1)$ essential submatrix of the following transition probability matrix:

$$\boldsymbol{M}_t = \begin{array}{c} 0 \\ 1 \\ \vdots \\ \vdots \\ k \\ \alpha \end{array}\left(\begin{array}{ccccc|c} q_t & p_t & 0 & \cdots & 0 & 0 \\ q_t & 0 & p_t & \cdots & 0 & 0 \\ \vdots & \vdots & \ddots & \ddots & & \vdots \\ \vdots & \vdots & & \ddots & \ddots & \vdots \\ q_t & 0 & \cdots & \cdots & 0 & p_t \\ \hline 0 & 0 & \cdots & \cdots & 0 & 1 \end{array}\right)_{(k+2) \times (k+2)} = \left(\begin{array}{c|c} \boldsymbol{N}_t & \boldsymbol{C}_t \\ \hline \boldsymbol{0} & 1 \end{array}\right). \tag{3.25}$$

Proof.   The longest success run in a sequence of two-state trials is related to the runs statistics $N_{n,k}$, $G_{n,k}$ and $M_{n,k}$ in the following simple way:

$$L_n(S) \le k \quad \text{if and only if} \quad N_{n,k+1} = G_{n,k+1} = M_{n,k+1} = 0.$$

Thus $P(L_n(S) \le k) = P(N_{n,k+1} = 0)$, and we can complete the proof by considering Eq. (3.8) for $P(N_{n,k+1} = x)$ with $x = 0$. Changing states $(0,0), (0,1), \cdots, (0,k)$ in this application of Eq. (3.8) to states $0, 1, 2, \cdots, k$, respectively, and combining all other states into the absorbing state α, we obtain

$$P(L_n(S) \le k) = P(N_{n,k+1} = 0) = \boldsymbol{\xi}_0(\prod_{t=1}^{n} \boldsymbol{M}_t)(1, \cdots, 1, 0)',$$

where $\boldsymbol{\xi}_0 = (1, 0, \cdots, 0)_{1 \times (k+2)} = (\boldsymbol{\xi} : 0)$.

With the notation $(1, \cdots, 1, 0)_{1 \times (k+2)} = (\boldsymbol{1} : 0)$, and since the product of the transition probability matrices has the form

$$\prod_{t=1}^{n} \boldsymbol{M}_t = \left(\begin{array}{c|c} \prod_{t=1}^{n} \boldsymbol{N}_t & \boldsymbol{C}_t(n) \\ \hline \boldsymbol{0} & 1 \end{array}\right),$$

the theorem follows immediately. $\square$

Corollary 3.1 *Given $1 \le k < n$, we have the recursive equation*

$$P(L_n(S) \le k) = q_n P(L_{n-1}(S) \le k) + \sum_{i=1}^{k} q_{n-i} \prod_{j=n-i+1}^{n} p_j P(L_{n-i-1}(S) \le k),$$

$$(3.26)$$

with $P(L_n(S) = 0) = \prod_{j=1}^{n} q_j$ and $P(L_n(S) \le n) \equiv 1$ for $k = n$.

Proof. It follows from the structure of the transition probability matrices M_t given by Eq. (3.25) that

(i) $M_t e_0' = q_t(1, \cdots, 1, 0)_{1 \times (k+2)}'$, and

(ii) $M_t e_i' = p_t e_{i-1}'$, for $i = 1, \cdots, k,$

where $e_i = (0, \cdots, 0, 1, 0, \cdots, 0)$ is a $1 \times (k+2)$ unit row vector having 1 at the coordinate associated with state i, for $i = 0, 1, \cdots, k$.

Since $(1, \cdots, 1, 0)' = \sum_{i=0}^{k} e_i'$, our result is a direct consequence of (i), (ii) and backward multiplications of Eq. (3.24). $\square$

Taking $p_t = q_t = 1/2$ for all $t = 1, \cdots, n$, and multiplying $P(L_n(S) \le k)$ by 2^n, Eq. (3.26) yields the recursive equation (3.21) for $A_n(k)$. Theorem 3.1 can also be extended to the longest failure run $L_n(F)$ and the longest run statistic $L_n = \max(L_n(S), L_n(F))$.

For *i.i.d.* cases and for large n, there are several outstanding results on the length of the longest success run. Rényi (1970), Csörgö (1979), Erdös and Rényi (1970) and Erdös and Révész (1975) show that, as $n \to \infty$,

$$\frac{L_n(S)}{\log_{1/p} n} \overset{a.s.}{\longrightarrow} 1.$$

This result is often referred to as the new law of large numbers.

In Chapter 5, we will develop a large deviation approximation for the probability of $L_n(S)$ under *i.i.d.* (and homogeneous Markov-dependent) trials:

$$P(L_n(S) \le k) \sim \exp\{-n\beta\},$$

where $\beta = -\log \lambda_{[1]}$ and $\lambda_{[1]}$ is the largest eigenvalue of the essential transition probability submatrix N_t (with constant p_t and q_t) given by Eq. (3.25).

3.7   Waiting-Time Distribution of a Success Run

Let $\Lambda = S \cdots S$ be the simple pattern of k consecutive successes, and define the random variable $W(\Lambda)$ as the waiting time for pattern Λ to occur, *i.e.*

$$W(\Lambda) = \inf\{n : X_{n-k+1} = X_{n-k+2} = \cdots = X_n = S\}.$$

For example, given $k = 3$, $W(\Lambda) = 6$ means that the pattern SSS occurs for the first time after six trials, as in $SFFSSS$. The distribution of $W(\Lambda)$ for Bernoulli trials is often referred to as the geometric distribution of order k (see Aki 1985 and Hirano 1986).

Theorem 3.2   *For a given pattern length $k \geq 1$ and a sequence of Bernoulli trials $\{X_t\}$, the distribution of $W(\Lambda)$ is given by*

$$P(W(\Lambda) = n) = \boldsymbol{\xi} \boldsymbol{N}^{n-1}(\Lambda)(\boldsymbol{I} - \boldsymbol{N}(\Lambda))\boldsymbol{1}', \tag{3.27}$$

where $\boldsymbol{\xi} = (1, 0, \cdots, 0)$ is a $1 \times k$ row vector, and $\boldsymbol{N}(\Lambda)$ is the $k \times k$ essential transition probability submatrix of

$$\boldsymbol{M}(\Lambda) = \begin{array}{c} 0 \\ 1 \\ \vdots \\ \vdots \\ k-1 \\ \alpha \end{array} \left(\begin{array}{cccccc|c} q & p & 0 & \cdots & 0 & 0 \\ q & 0 & p & \cdots & 0 & 0 \\ \vdots & & \ddots & \ddots & & \vdots \\ \vdots & & & \ddots & \ddots & \vdots \\ q & 0 & \cdots & \cdots & 0 & p \\ 0 & 0 & \cdots & \cdots & 0 & 1 \end{array} \right)_{(k+1)\times(k+1)} = \left(\begin{array}{c|c} \boldsymbol{N}(\Lambda) & \boldsymbol{C} \\ \hline \boldsymbol{0} & 1 \end{array} \right).$$
$$\tag{3.28}$$

It further follows that the probability generating function of $W(\Lambda)$ is given by

$$\varphi_W(s) = \frac{p^k s^k (1 - ps)}{1 - s + q p^k s^{k+1}}. \tag{3.29}$$

Proof.   Given Λ, n, and $k \leq n$, it follows from the definition of $W(\Lambda)$ and $N_{n,k}$ that these two random variables have the following relationship:

$$W(\Lambda) \leq n \quad \text{if and only if} \quad N_{n,k} \geq 1, \ \forall n \geq k.$$

Hence $P(W(\Lambda) \leq n) = P(N_{n,k} \geq 1)$, and

$$\begin{aligned} P(W(\Lambda) = n) &= P(N_{n,k} \geq 1) - P(N_{n-1,k} \geq 1), \\ &= P(N_{n-1,k} = 0) - P(N_{n,k} = 0). \end{aligned} \tag{3.30}$$

Since we only need the general result for $P(N_{n,k} = 0)$ to complete the proof, we can, as in the previous section on the longest success run, replace the states $(0,0), \cdots, (0, k-1)$ defined in Section 3.2 by the states $0, 1, \cdots, k-1$, and combine all other states into an absorbing state α. Under this reduced state space, the transition probability matrix $\boldsymbol{M}(N_{n,k})$ is simplified to $\boldsymbol{M}(\Lambda)$. Equation (3.27) of Theorem 3.2 is then an immediate consequence of Eqs. (3.8), (3.30) and Theorem 3.1.

Note that, for $0 \le i \le k-1$,

$$\boldsymbol{e}_i \boldsymbol{N}(\Lambda) = q\boldsymbol{e}_0 + p\boldsymbol{e}_{i+1}.$$

Given $\boldsymbol{\xi} = \boldsymbol{e}_0$, using the above forward-multiplication result yields the following recursive equation:

$$P(W(\Lambda) = n) = \boldsymbol{\xi} \boldsymbol{N}^{n-1}(\Lambda)(\boldsymbol{I} - \boldsymbol{N}(\Lambda))\boldsymbol{1}' = \sum_{i=1}^{k} qp^{i-1} P(W(\Lambda) = n - i).$$

$$(3.31)$$

The above recursive equation and the boundary condition $P(W(\Lambda) = k) = p^k$ lead to the following recursive equation for the probability generating function $\varphi_W(s)$:

$$\varphi_W(s) = s^k p^k + \sum_{i=1}^{k} qp^{i-1} s^i \varphi_W(s).$$

Summing the finite power series yields the explicit result for $\varphi_W(s)$ given in Eq. (3.29), a result which was first derived by Feller (1968) using renewal theory. $\square$

Let us define the matrices

$$\boldsymbol{A} = \begin{pmatrix} q & p & 0 & \cdots & 0 \\ q & 0 & p & & \\ \vdots & & \ddots & \ddots & \\ q & & & 0 & p \\ q & 0 & \cdots & & 0 \end{pmatrix}_{k \times k}, \quad \boldsymbol{B} = \begin{pmatrix} 0 & 0 & 0 & \cdots & 0 \\ 0 & 0 & 0 & & \\ \vdots & & \ddots & \ddots & \\ 0 & & & 0 & 0 \\ p & 0 & \cdots & & 0 \end{pmatrix}_{k \times k},$$

$$\boldsymbol{A}^* = \begin{pmatrix} 1 \end{pmatrix}_{1 \times 1}, \text{ and } \boldsymbol{B}^* = \begin{pmatrix} 0 & 0 & \cdots & 0 & p \end{pmatrix}'_{k \times 1},$$

and let $W(m, \Lambda)$ be the waiting time for the m-th run of k consecutive successes (under non-overlap counting). Similar to the above development for $W(\Lambda)$, the distribution of the random variable $W(m, \Lambda)$ can also be obtained using Eq. (3.27) by replacing the transition probability matrix $M(\Lambda)$ by

$$
M(m, \Lambda) = \begin{pmatrix} A & B & & & O \\ & \ddots & \ddots & & \\ & & A & B & \\ O & & & A & B^* \\ & & & & A^* \end{pmatrix}_{(mk+1) \times (mk+1)}.
$$

Note that the transition probability matrix $M(\Lambda)$ in Eq. (3.28) is the special case of $M(m, \Lambda)$ with $m = 1$. Since $W(m, \Lambda) = \sum_{i=1}^{m} W_i(\Lambda)$, where $W_i(\Lambda)$ represents the waiting time from the $(i-1)$-th occurrence until the i-th occurrence of pattern Λ, and since the random variables $W_i(\Lambda)$ are *i.i.d.*, it follows from Eq. (3.29) that the probability generating function of $W(m, \Lambda)$ is

$$
\varphi_{W(m,\Lambda)}(s) = \varphi_W^m(s) = \left(\frac{p^k s^k (1 - ps)}{1 - s + qp^k s^{k+1}} \right)^m. \tag{3.32}
$$

The probability generating function $\varphi_W(s)$ always exists for all $|s| \leq 1$. This follows from its definition and the fact that

$$
|\varphi_W(s)| \leq \sum_{n=1}^{\infty} |s^n| P(W = n) \leq \sum_{n=1}^{\infty} P(W = n) = 1.
$$

However, the probability generating function $\varphi_W(s)$ may exist beyond the region $|s| \leq 1$. The exact region varies from problem to problem. We shall return to discuss the largest region of existence for $\varphi_W(s)$ in Section 5.7.

The distribution of $W(m, \Lambda)$ can also be obtained through the equation

$$
P(W(m, \Lambda) = n) = \frac{1}{n!} \frac{d^n}{ds^n} \varphi_{W(m,\Lambda)}(s)|_{s=0},
$$

an approach which may be most easily accomplished using symbolic manipulation software (*e.g.* MAPLE or MATLAB). Further treatment of waiting-time distributions is given in Chapter 5 for both simple and compound patterns under *i.i.d.* as well as Markov-dependent multi-state trials.

3.8 Numerical Examples

Before studying more complex runs statistics, in this section we provide some numerical results for the runs statistics and waiting times described in previous sections in order to illustrate the theoretical results. Given the transition probability matrix (or matrices) of the imbedded Markov chain $\{Y_t\}$, in general we need only two kinds of formulae, in the forms of Eqs. (3.8) and (3.27), for evaluating the distributions of $X_n(\Lambda)$ and $W(\Lambda)$, respectively. The formulae are simple and computationally efficient, suitable even for very large n. The numerical results presented here can also serve the purpose of checking one's programming computations. In all the examples considered, the computational time for obtaining each distribution is minimal, a fraction of a second on a current PC.

Table 3.1 gives the exact distributions and means of the random variables $E_{15,2}$, $G_{15,2}$, $N_{15,2}$, $M_{15,2}$ and $L_{15}(S)$ under the assumption that $\{X_t\}_{t=1}^{15}$ is a sequence of independent two-state trials with probabilities $p_t = 1/(t+1)$ for $t = 1, 2, \cdots, 15$.

Table 3.1 The exact distributions and means of five runs statistics with $n = 15$ and $p_t = 1/(t+1)$.

$x \backslash r.v.$	$E_{15,2}$	$G_{15,2}$	$N_{15,2}$	$M_{15,2}$	$L_{15}(S)$
0	.73200	.67163	.67163	.67163	.06250
1	.24771	.30120	.29046	.24125	.60913
2	.01976	.02646	.03602	.06881	.26135
3	.00051	.00070	.00184	.01516	.05531
4	3.99e-07	5.27e-08	.00004	.00270	.00991
5	5.08e-09	5.69e-09	4.85e-07	.00040	.00156
6			2.28e-09	.00005	.00021
7			3.10e-12	5.68e-06	.00003
8				5.54e-07	2.88e-06
9				4.81e-08	2.85e-07
10				3.71e-09	2.58e-08
11				2.55e-10	2.14e-09
12+				1.58e-11	1.64e-10
Mean	.28879	.35625	.36821	.43750	1.34668

Table 3.2 gives the waiting-time distributions of the first success run
of length k, for various values of k and state probabilities p_t. Additional
numerical results on waiting-time distributions will be provided in Chapters
5 and 7.

Table 3.2   Some waiting-time distributions of the first success run of length k.

n	$p = .9$		$p_t = t/(t+1)$		$p_t = 1 - 1/2^t$		
	$k = 2$	$k = 3$	$k = 2$	$k = 3$	$k = 2$	$k = 3$	$k = 5$
2	.8100		.3333		.3750		
3	.0810	.7290	.2500	.2500	.3281	.3281	
4	.0810	.0729	.2000	.2000	.2051	.3076	
5	.0154	.0729	.1111	.1667	.0710	.1987	.2980
6	.0088	.0729	.0595	.1429	.0177	.1118	.2933
7	.0023	.0198	.0271	.0937	.0028	.0397	.1940
8	.0010	.0144	.0117	.0611	.0003	.0111	.1104
9	.0003	.0091	.0046	.0383		.0026	.0588
10	.0001	.0038	.0017	.0219		.0004	.0303
11		.0024	.0006	.0122		.0001	.0108
12		.0013	.0002	.0066			.0032
13		.0007	.0001	.0034			.0008
14		.0004		.0017			.0002
15$^+$		.0002		.0008			

3.9   Number of Successes in Success Runs of Length Greater Than or Equal to k

Let $S_{n,k}$ be the total number of successes in success runs of length longer
than or equal to k, for $k = 1, \cdots, n$. It can be written as

$$S_{n,k} = \sum_{i=k}^{n} i R_n(i), \tag{3.33}$$

where $R_n(i)$, for $i = k, \cdots, n$, is the number of success runs of length exactly
equal to i in a sequence $\{X_t\}$. For $k = 1$, Eq. (3.33) is equivalent to the

total number of successes in the sequence $\{X_t\}$; *i.e.*

$$S_{n,1} = \sum_{i=1}^{n} I_{X_i},$$

where $I_{X_i} = 1$ if the i-th trial is a success and zero otherwise. If $\{X_t\}$ is a sequence of Bernoulli trials, then $S_{n,1}$ has binomial and normal distributions as exact and limiting distributions, respectively. More generally, for the case where $\{X_t\}$ is a sequence of Markov-dependent random variables, the statistic $S_{n,k}$ also has a normal limiting distribution, as determined by Nagaev (1957) for $k = 1$ and by Fu, Lou, Bai and Li (2002) for $k \geq 2$. In this section, we only study the exact distribution of $S_{n,k}$ with $k \geq 2$.

Let L_j, for $j \geq 2$, be the length of the success run located between the $(j-1)$-th and j-th failures in the sequence $\{X_t\}$, with $L_1 = 0$ if the first trial is a failure and $L_1 = l$ if the first l trials are successes and the $(l+1)$-th trial is a failure. For a given time index t, let m_t be the number of failures in the subsequence $X_1, X_2, \cdots, X_t$, and let $L_t^\star$ represent the number of successes that occur after the m_t-th failure in this subsequence. Note that $0 \leq L_t^\star \leq t$ and $0 \leq L_t^\star \leq L_{m_t+1}$. Moreover, $S_{t,k}$, as defined by Eq. (3.33), can also be written as

$$S_{t,k} = \sum_{j=1}^{m_t} L_j(k) + L_t^\star(k), \tag{3.34}$$

where

$$L_j(k) = L_j I_{\{L_j \geq k\}}, \text{ and } L_j^\star(k) = L_j^\star I_{\{L_j^\star \geq k\}}. \tag{3.35}$$

Here $I_{\{L_j \geq k\}}$ is an indicator function equal to one if $L_j \geq k$ and zero otherwise ($I_{\{L_j^\star \geq k\}}$ is defined similarly).

To capture the relevant information in the subsequence $X_1, X_2, \cdots, X_t$, we define a new sequence of random variables in the form of the two-component vector

$$Y_t = (S_{t,k}, E_t(k)), \ t = 1, \cdots, n, \tag{3.36}$$

where $S_{t,k}$ indicates the total number of successes in success runs of length greater than or equal to k in the first t trials, and $E_t(k)$ is the ending-block random variable given by

$$E_t(k) = L_t^\star(1 - I_{\{L_t^\star \geq k\}}) + k^+ I_{\{L_t^\star \geq k\}}.$$

In this expression, k^+ is a symbol representing the state where $L_t^\star$ is greater than or equal to k.

The ending block $E_t(k)$ represents the length of the success run counting backward from the t-th trial, with $E_t(k) = 0$ if the t-th trial is a failure and $E_t(k) = k^+$ if the length is greater than or equal to k. More specifically, consider that from the m_t-th (or most recent) failure F to the end of the subsequence $X_1, X_2, \cdots, X_t$, we can only have the following possible outcomes: $\{F, FS, \cdots, FS \cdots S, \text{ or } S \cdots S \text{ if } m_t = 0\}$; the random variable $E_t(k)$ is equal to the number of successes in these outcomes if the number is less than k, and $E_t(k) = k^+$ if it is equal to or greater than k. This ending block of the first t trials provides essential information about the transition probabilities from Y_t to Y_{t+1}.

We define the state space

$$\Omega = \{(u, v) : u = 0, k, \cdots, n - 1, n; \ v = 0, 1, \cdots, k - 1, k^+\} \qquad (3.37)$$

with size $d = card(\Omega) = (n - k + 2)(k + 1)$, and consider here the case where the sequence $\{X_t\}$ is a homogeneous Markov chain with transition probabilities p_{FF}, p_{FS}, p_{SF}, and p_{SS}. In our counting procedure, the sequence of random vectors $Y_t = (S_{t,k}, E_t(k))$, $t = 1, 2, \cdots, n$, defined on Ω obeys the following rules:

(i) Given $Y_{t-1} = (x, 0)$, then $Y_t = (x, 0)$ with probability p_{FF} if the outcome of the t-th trial is F, and $Y_t = (x, 1)$ with probability p_{FS} if the outcome of the t-th trial is S.

(ii) Given $Y_{t-1} = (x, y)$ for $1 \leq y \leq k - 2$, then $Y_t = (x, 0)$ with probability p_{SF} if the outcome of the t-th trial is F, and $Y_t = (x, y + 1)$ with probability p_{SS} if the outcome of the t-th trial is S.

(iii) Given $Y_{t-1} = (x, k - 1)$, then $Y_t = (x, 0)$ with probability p_{SF} if the outcome of the t-th trial is F, and $Y_t = (x + k, k^+)$ with probability p_{SS} if the outcome of the t-th trial is S.

(iv) Given $Y_{t-1} = (x, k^+)$, then $Y_t = (x, 0)$ with probability p_{SF} if the outcome of the t-th trial is F, and $Y_t = (x + 1, k^+)$ with probability p_{SS} if the outcome of the t-th trial is S.

In view of our construction, the sequence $\{Y_t = (S_{t,k}, E_t(k)) : t = 1, 2, \cdots, n\}$ forms a homogeneous Markov chain with transition probability matrix

$$M = \left(p_{(x,y),(u,v)}\right)_{d \times d},$$

where the transition probabilities $p_{(x,y),(u,v)}$, under lexicographical ordering of the states $(\cdot,\cdot)$, can be specified explicitly as follows. Given $(x,y) \in \Omega$,

$$p_{(x,y),(u,v)} = \begin{cases} p_{FF} & \text{if } y = v = 0 \text{ and } u = x \\ p_{FS} & \text{if } y = 0, v = 1, \text{ and } u = x \\ p_{SF} & \text{if } y \neq 0, v = 0, \text{ and } u = x \\ p_{SS} & \text{if } 1 \leq y \leq k-2, v = y+1, \text{ and } u = x, \\ & \quad \text{or if } y = k-1, v = k^{+}, \text{ and } u = x+k, \\ & \quad \text{or if } y = k^{+}, v = k^{+}, \text{ and } u = x+1 \\ 1 & \text{if } y = v \text{ and } u = x = n \\ 0 & \text{otherwise.} \end{cases} \tag{3.38}$$

Hence the random variable $S_{n,k}$ is finite Markov chain imbeddable, and the exact probabilities can be obtained from

$$P(S_{n,k} = x) = \boldsymbol{\xi}_0 \boldsymbol{M}^{n-1} \boldsymbol{U}'(C_x), \quad x = 0, k, \cdots, n, \tag{3.39}$$

where the $1 \times d$ row vector $\boldsymbol{\xi}_0 = (q_0, p_0, 0, \cdots, 0)$ is the initial distribution of Y_1, the partition $\{C_x\}$ is defined as

$$C_x = \{(x,y) : y = 0, 1, \cdots, k-1, k^{+}\}, \quad x = 0, k, \cdots, n,$$

and $\boldsymbol{U}'(C_x)$ is the transpose of the $1 \times d$ row vector $\boldsymbol{U}(C_x) = (0, \cdots, 0, 1, \cdots, 1, 0, \cdots, 0)$ with ones at the coordinates corresponding to states in C_x.

To gain further insight on the effects of the various parameters, the distributions of $S_{n,k}$ for a few representative cases are shown graphically in Figure 3.1 with $n = 15$, 30, and 60, where the initial distribution $\boldsymbol{\xi}_0 = (1, 0, \cdots, 0)$ is assumed. When p_{SS} is small (*e.g.* $p_{SS} = 0.2$), the parameters' effects on the distribution are less pronounced, and hence in Figure 3.1, only cases with large values of p_{SS} ($= 0.8$) are presented. For purposes of comparison, the expectations are also included.

For fixed n, the effect of k and p_{FS} can be summarized as follows. For small k ($k = 2$), the distribution of $S_{n,k}$ becomes smooth and bell-shaped as n increases (from Figure 3.1(a) to (d) to (g)), and this tendency is amplified at larger values of p_{FS}. As k increases, the distributions drift away from a normal shape and become highly skewed to the right (for example, from Figure 3.1(d) to (e) to (f)). The distribution of $S_{n,k}$ can be approximated by a normal distribution only when k is much smaller than n, and normal approximations should be used with caution. Further details of the limiting distribution of $S_{n,k}$ are given in Fu, Lou, Bai and Li (2002).

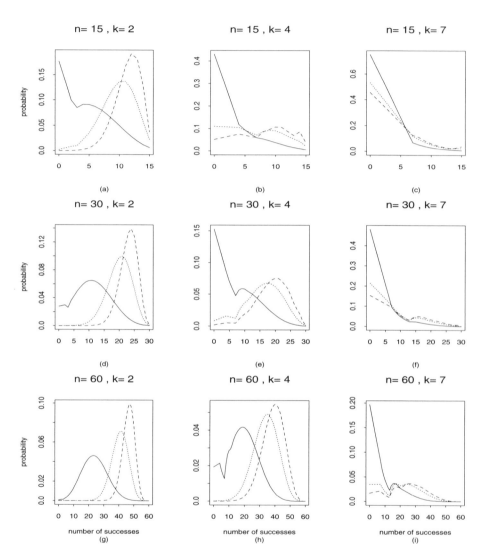

Fig. 3.1   Representative distributions of $S_{n,k}$ with $n = 15, 30, 60$, $k = 2, 4, 7$, $p_{SS} = 0.8$ and $p_{FS} = 0.15$(solid line), 0.5(dotted line), 0.85(dashed line). The means of the distributions are, in ascending order of p_{FS}, as follows: $ES_{n,k} =$ (a) 5.27, 9.76, 11.44, (b) 4.09, 7.65, 8.98, (c) 2.30, 4.43, 5.24, (d) 11.44, 20.05, 23.09, (e) 9.35, 16.43, 18.93, (f) 6.01, 10.61, 12.25, (g) 23.78, 40.62, 46.41, (h) 19.89, 33.98, 38.83, (i) 13.43, 22.97, 26.25.

Chapter 4

Runs and Patterns in Multi-State Trials

4.1 Introduction

In Chapter 3, we discussed key ideas of the finite Markov chain imbedding technique to obtain the exact distributions for the numbers of success runs and patterns in a sequence of two-state trials. The main goal of this chapter is to extend the finite Markov chain imbedding technique to study the number of runs and patterns in a sequence of multi-state trials. It might seem that, in principle, the extension should be somewhat straightforward and should require only minor modifications. However, this is not the case, especially when the pattern is complex and the sequence $\{X_t\}$ consists of Markov-dependent multi-state trials. The main difficulties are due to the complexity of constructing a proper finite Markov chain associated with the random variable $X_n(\Lambda)$, especially in the process of obtaining the transition probabilities. In order to overcome these difficulties, we introduce the *forward and backward principle*. Our focus in this chapter will be on using the forward and backward principle to obtain the distributions of simple and compound patterns. In fact, the forward and backward principle plays an indispensable role in constructing the imbedded Markov chain for almost every application covered by this book.

4.2 Forward and Backward Principle with Non-Overlap Counting

Let us start with the simple case that $\{X_t\}_{t=1}^n$ is a sequence of *i.i.d.* multi-state trials. Each trial has m $(m \geq 2)$ possible outcomes (states or symbols), labeled $\mathcal{S} = \{b_1, \cdots, b_m\}$ and occurring with probabilities $p_1, \cdots, p_m$, respectively. Denote by $X_n(\Lambda)$ the number of *non-overlapping* simple patterns Λ in the sequence $\{X_t\}$. First, we would like to introduce the forward and backward principle for the finite Markov chain imbedding technique, a principle which will guide the construction of an imbedded Markov chain $\{Y_t\}$ and the determination of its transition probability matrices. For convenience of discussion, the forward and backward principle is introduced via the following example.

Example 4.1   Let us consider a simple pattern $\Lambda = b_1 b_1 b_2$ in a sequence of n three-state trials ($\mathcal{S} = \{b_1, b_2, b_3\}$).

(i) Decompose the pattern $\Lambda = b_1 b_1 b_2$ into a set of *sequential subpatterns* $\mathcal{S}(\Lambda) = \{b_1, b_1 b_1, b_1 b_1 b_2\}$. Define

$$\mathcal{E} = \mathcal{S} \cup \mathcal{S}(\Lambda) = \{b_1, b_2, b_3, b_1 b_1, b_1 b_1 b_2\} \tag{4.1}$$

as a set of ending blocks induced by the pattern $b_1 b_1 b_2$ with respect to the sequence of trials $\{X_t\}$.

(ii) Let $\omega = (x_1, \cdots, x_n)$ be a realization of a sequence of n three-state trials. We define a state space

$$\Omega = \{(u, v) : u = 0, 1, \cdots, [n/3], v \in \mathcal{E}\} \cup \{\emptyset\} - \{(0, b_1 b_1 b_2)\} \tag{4.2}$$

and a Markov chain

$$\{Y_t = (X_t(\Lambda), E_t), \ t = 1, 2, \cdots, n\} \tag{4.3}$$

operating on ω as

$$Y_t(\omega) = (u, v), \ \text{for } t = 1, \cdots, n,$$

where

$$
\begin{aligned}
u &= X_t(\Lambda)(\omega) = \text{the total number of non-overlapping patterns } \Lambda \\
  &\quad \text{in the first } t \text{ trials, counting } forward \text{ from the first trial to} \\
  &\quad \text{the } t\text{-th trial, and} \\
v &= E_t(\omega) = \text{the longest ending block in } \mathcal{E}, \text{ counting } backward \\
  &\quad \text{from } X_t.
\end{aligned}
$$

The definitions of u and v for the sequence of the first t trials are graphically illustrated in Figure 4.1.

To make the imbedded Markov chain $\{Y_t\}$ and the concept of the longest ending block more transparent, let us consider the following realization,

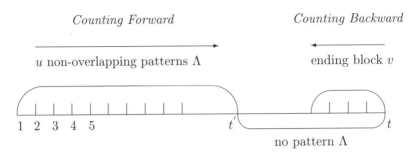

Fig. 4.1   Diagram illustrating the forward and backward counting procedure. Note that when $t' = t$ and the u-th pattern occurred at the t-th trial, then, under non-overlap counting, the longest possible ending block is the pattern itself ($v = \Lambda$).

$\omega = (b_3 b_1 b_2 b_1 b_1 b_2 b_1)$, of a sequence of seven three-state trials. Applying the forward and backward principle, the corresponding realization of the imbedded Markov chain $\{Y_t\}$ on ω is given by $\{Y_1(\omega) = (0, b_3), Y_2(\omega) = (0, b_1), Y_3(\omega) = (0, b_2), Y_4(\omega) = (0, b_1), Y_5(\omega) = (0, b_1 b_1), Y_6(\omega) = (1, b_1 b_1 b_2)$ and $Y_7(\omega) = (1, b_1)\}$. Note that for every given ω, the realization of the imbedded Markov chain $Y_t(\omega) = (u, v)$ is uniquely determined by the above procedures (i) and (ii) under non-overlap counting. In plain words, the ending block v represents the status of forming the next pattern Λ for the subsequence $\{x_1, \cdots, x_t\}$ containing u complete patterns.

(iii) The imbedded Markov chain $\{Y_t\}$ is homogeneous and its transition probability matrix $\boldsymbol{M} = (p_{(x,z),(u,v)})$ may be determined as follows. For example, given $Y_5(\omega) = (0, b_1 b_1)$, since X_6 can only be one of the three possible outcomes b_1, b_2 and b_3, the forward and backward counting procedure yields

$$
\begin{array}{ccl}
Y_5(\omega) & \rightarrow & Y_6(\omega) \\[4pt]
(0, b_1 b_1) & \rightarrow & \left\{
\begin{array}{ll}
(0, b_1 b_1) & \text{if } X_6 = b_1 \text{ (with probability } p_1) \\
(1, b_1 b_1 b_2) & \text{if } X_6 = b_2 \text{ (with probability } p_2) \\
(0, b_3) & \text{if } X_6 = b_3 \text{ (with probability } p_3),
\end{array}
\right.
\end{array} \tag{4.4}
$$

and $Y_5(\omega)$ goes to any other state with probability zero. In this manner, all of the transition probabilities $P(Y_t = (u, v) | Y_{t-1} = (x, z))$ can be obtained. The dummy state $\emptyset$ will be added as an initial state with $P(Y_0 = \emptyset) = 1$ and with transition probabilities $P(Y_1 = b_i | Y_0 = \emptyset) = p_i$ for $i = 1, 2, 3$. Note that the state $(0, \Lambda = b_1 b_1 b_2)$ was deleted from the state space, since whenever the ending block v equals Λ there must be at least one occurrence of the pattern in the sequence (*i.e.* $u \geq 1$ if $v = \Lambda$).

(iv) Given n, we have the following partition on the state space Ω:

$$\{C_\emptyset = [\emptyset], C_0 = [(0,b_1),(0,b_2),(0,b_3),(0,b_1b_1)],$$
$$\text{and } C_x = [(x,v), v \in \mathcal{E}], x = 1, \cdots, [n/3]\}. \tag{4.5}$$

For $n = 5$ and the initial probability $P(Y_0 = \emptyset) \equiv 1$, it follows from the above procedures (i) to (iv) that the imbedded Markov chain $\{Y_t\}_{t=0}^5$ is defined on the state space $\Omega = \{\emptyset, (0,b_1), (0,b_2), (0,b_3), (0,b_1b_1), (1,b_1b_1b_2), (1,b_1), (1,b_2), (1,b_3), (1,b_1b_1)\}$ with transition probability matrix

$$\boldsymbol{M} = \begin{array}{c} \emptyset \\ (0,b_1) \\ (0,b_2) \\ (0,b_3) \\ (0,b_1b_1) \\ (1,b_1b_1b_2) \\ (1,b_1) \\ (1,b_2) \\ (1,b_3) \\ (1,b_1b_1) \end{array} \left(\begin{array}{ccccc|ccccc} 0 & p_1 & p_2 & p_3 & 0 & 0 & 0 & 0 & 0 & 0 \\ 0 & 0 & p_2 & p_3 & p_1 & 0 & 0 & 0 & 0 & 0 \\ 0 & p_1 & p_2 & p_3 & 0 & 0 & 0 & 0 & 0 & 0 \\ 0 & p_1 & p_2 & p_3 & 0 & 0 & 0 & 0 & 0 & 0 \\ 0 & 0 & 0 & p_3 & p_1 & p_2 & 0 & 0 & 0 & 0 \\ 0 & 0 & 0 & 0 & 0 & 0 & p_1 & p_2 & p_3 & 0 \\ 0 & 0 & 0 & 0 & 0 & 0 & 0 & p_2 & p_3 & p_1 \\ 0 & 0 & 0 & 0 & 0 & 0 & p_1 & p_2 & p_3 & 0 \\ 0 & 0 & 0 & 0 & 0 & 0 & p_1 & p_2 & p_3 & 0 \\ 0 & 0 & 0 & 0 & 0 & 0 & 0 & 0 & 0 & 1 \end{array} \right). \tag{4.6}$$

The probabilities $P(X_5(\Lambda) = x) = \boldsymbol{\xi}_0 \boldsymbol{M}^5 \boldsymbol{U}'(C_x)$, $x = 0,1$, can then be easily computed. $\diamond$

To demonstrate the applicability of the forward and backward principle to compound patterns, let us consider the following example.

Example 4.2   Given $n = 4$ and a compound pattern $\Lambda = \Lambda_1 \cup \Lambda_2$ consisting of the union of two distinct simple patterns $\Lambda_1 = b_1b_2$ and $\Lambda_2 = b_3b_1$, we are interested in finding the distribution of the random variable $X_4(\Lambda)$, the number of occurrences of either Λ_1 or Λ_2 in a sequence of four *i.i.d.* three-state trials. Proceeding as in the previous example, one obtains the imbedded Markov chain $\{Y_t\}_0^4$ defined on the state space

$$\begin{aligned} \Omega &= \{\emptyset, (0,b_1), (0,b_2), (0,b_3), (1,b_1b_2), (1,b_3b_1), (1,b_1), \\ &\quad (1,b_2), (1,b_3), (2,b_1b_2), (2,b_3b_1)\} \end{aligned} \tag{4.7}$$

with transition probability matrix

$$
\boldsymbol{M} =
\begin{array}{c}
\emptyset \\
(0,b_1) \\
(0,b_2) \\
(0,b_3) \\
(1,b_1b_2) \\
(1,b_3b_1) \\
(1,b_1) \\
(1,b_2) \\
(1,b_3) \\
(2,b_1b_2) \\
(2,b_3b_1)
\end{array}
\left(
\begin{array}{ccccccccccc}
0 & p_1 & p_2 & p_3 & 0 & 0 & 0 & 0 & 0 & 0 & 0 \\
0 & p_1 & 0 & p_3 & p_2 & 0 & 0 & 0 & 0 & 0 & 0 \\
0 & p_1 & p_2 & p_3 & 0 & 0 & 0 & 0 & 0 & 0 & 0 \\
0 & 0 & p_2 & p_3 & 0 & p_1 & 0 & 0 & 0 & 0 & 0 \\
0 & 0 & 0 & 0 & p_1 & 0 & 0 & p_2 & p_3 & 0 & 0 \\
0 & 0 & 0 & 0 & 0 & 0 & p_1 & p_2 & p_3 & 0 & 0 \\
0 & 0 & 0 & 0 & 0 & 0 & p_1 & 0 & p_3 & p_2 & 0 \\
0 & 0 & 0 & 0 & 0 & 0 & p_1 & p_2 & p_3 & 0 & 0 \\
0 & 0 & 0 & 0 & 0 & 0 & 0 & p_2 & p_3 & 0 & p_1 \\
0 & 0 & 0 & 0 & 0 & 0 & 0 & 0 & 0 & 1 & 0 \\
0 & 0 & 0 & 0 & 0 & 0 & 0 & 0 & 0 & 0 & 1
\end{array}
\right).
$$

$$(4.8)$$

The probabilities $P(X_4(\Lambda) = x) = \boldsymbol{\xi}_0 \boldsymbol{M}^4 \boldsymbol{U}'(C_x)$, $x = 0, 1, 2$, can again be computed with ease. $\diamond$

The method can also be extended, with simple modifications, to the case where $\{X_t\}$ is a sequence of Markov-dependent multi-state trials.

Example 4.3   Let us return to Example 4.1, but consider here that $\{X_t\}$ is a sequence of Markov-dependent three-state trials with transition probability matrix

$$
\boldsymbol{A} =
\left(
\begin{array}{ccc}
p_{11} & p_{12} & p_{13} \\
p_{21} & p_{22} & p_{23} \\
p_{31} & p_{32} & p_{33}
\end{array}
\right).
$$

Our goal is to determine the distribution of the pattern $\Lambda = b_1 b_1 b_2$ in a sequence of five trials. Analogous to Example 4.1, the transition probabilities of the imbedded Markov chain can be obtained for each state by the following argument. Given $Y_3 = (0, b_1 b_1)$, for example, we have

$$
\begin{array}{ccl}
Y_3 & \rightarrow & Y_4 \\
& & \left\{
\begin{array}{ll}
(0, b_1 b_1) & \text{if } X_4 = b_1 \text{ (with probability } p_{11}) \\
(1, b_1 b_1 b_2) & \text{if } X_4 = b_2 \text{ (with probability } p_{12}) \\
(0, b_3) & \text{if } X_4 = b_3 \text{ (with probability } p_{13}).
\end{array}
\right.
\end{array}
\quad (4.9)
$$

$$
(0, b_1 b_1) \rightarrow
$$

Equations (4.4) and (4.9) are equivalent except that the probabilities p_1, p_2 and p_3 are replaced by p_{11}, p_{12}, and p_{13}, respectively. Hence, the imbedded Markov chain $\{Y_t\}$ is defined here on the same state space Ω and has the transition probability matrix

$$
M = \begin{array}{c}
\emptyset \\
(0,b_1) \\
(0,b_2) \\
(0,b_3) \\
(0,b_1b_1) \\
(1,b_1b_1b_2) \\
(1,b_1) \\
(1,b_2) \\
(1,b_3) \\
(1,b_1b_1)
\end{array}
\left(
\begin{array}{ccccc|ccccc}
0 & p_1 & p_2 & p_3 & 0 & 0 & 0 & 0 & 0 & 0 \\
0 & 0 & p_{12} & p_{13} & p_{11} & 0 & 0 & 0 & 0 & 0 \\
0 & p_{21} & p_{22} & p_{23} & 0 & 0 & 0 & 0 & 0 & 0 \\
0 & p_{31} & p_{32} & p_{33} & 0 & 0 & 0 & 0 & 0 & 0 \\
0 & 0 & 0 & p_{13} & p_{11} & p_{12} & 0 & 0 & 0 & 0 \\ \hline
0 & 0 & 0 & 0 & 0 & 0 & p_{21} & p_{22} & p_{23} & 0 \\
0 & 0 & 0 & 0 & 0 & 0 & 0 & p_{12} & p_{13} & p_{11} \\
0 & 0 & 0 & 0 & 0 & 0 & p_{21} & p_{22} & p_{23} & 0 \\
0 & 0 & 0 & 0 & 0 & 0 & p_{31} & p_{32} & p_{33} & 0 \\
0 & 0 & 0 & 0 & 0 & 0 & 0 & 0 & 0 & 1
\end{array}
\right),
$$

$$(4.10)$$

where the transition probabilities $P(Y_t(u,v)|Y_{t-1} = (x,z))$ are obtained as illustrated through Eq. (4.9). Note that the transition probability matrix in Eq. (4.10) has the exact same form as the matrix in Eq. (4.6) for the i.i.d. case. ◇

In view of the state transitions outlined by Eqs. (4.4) and (4.9), leading to the transition probability matrices of Examples 4.1 to 4.3 given in Eqs. (4.6), (4.8), and (4.10), we define the following notation: given $Y_{t-1} = (x,z) \in \Omega$ and $X_t = j \in S$,

$$(u,v) \equiv < (x,z), j >_\Omega, \qquad (4.11)$$

where the state $(u,v) \in \Omega$ is the result of forward and backward (non-overlap) counting when an additional outcome $X_t = j$ is included. For every $(x,z) \in \Omega$, we also define $L(z) \in S$ to be the last element of the ending block z. Then for the general case, the transition probabilities of the imbedded Markov chain $\{Y_t\}$ are specified by the following equation:

$$
P(Y_t = (u,v)|Y_{t-1} = (x,z)) = \begin{cases} p_{ij} & \text{if } X_t = j \in S, L(z) = i \\ & \text{and } (u,v) = < (x,z), j >_\Omega \\ 0 & \text{otherwise}, \end{cases}
$$

$$(4.12)$$

where p_{ij} are the transition probabilities of the Markov chain $\{X_t\}$. If $\{X_t\}$ is a sequence of i.i.d. multi-state trials, then Eq. (4.12) becomes

$$
P(Y_t = (u,v)|Y_{t-1} = (x,z)) = \begin{cases} p_j & \text{if } X_t = j \in S \\ & \text{and } (u,v) = < (x,z), j >_\Omega \quad (4.13) \\ 0 & \text{otherwise}. \end{cases}
$$

Theorem 4.1 *Assuming that $\{X_t\}$ is a homogeneous Markov chain with transition probability matrix $\boldsymbol{A} = (p_{ij})_{m \times m}$, and $\Lambda = \cup_{i=1}^{l} \Lambda_i$ is a compound pattern generated by l distinct simple patterns Λ_i having the same length k, then the imbedded Markov chain $\{Y_t = (X_t(\Lambda), E_t), t = 1, 2, \cdots, n\}$ corresponding to the random variable $X_n(\Lambda)$*

(i) is defined on the state space

$$\begin{aligned}
\Omega &= \{\emptyset\} \cup \{(x, z) : x = 0, 1, \cdots, [n/k], z \in \mathcal{E}\} \qquad (4.14) \\
&\quad -\{(0, \Lambda_i) : i = 1, \cdots, l\} - \{([n/k], z) : k[n/k] + z(k) > n\},
\end{aligned}$$

where $\mathcal{E} = \mathcal{S} \cup_{i=1}^{l} \mathcal{S}(\Lambda_i)$ and $z(k) \equiv$ (length of z) mod k,

(ii) has the transition probability matrix

$$\boldsymbol{M} = \left(p_{(x,z)(u,v)}\right)_{d \times d}, \qquad (4.15)$$

where the transition probabilities are given by

$$
p_{(x,z)(u,v)} = \begin{cases}
p_j & \text{if } (x, z) = \emptyset, u = 0, v = j, \text{ for all } j \in \mathcal{S} \\
p_{ij} & \text{if } (u, v) = <(x, z), j>_{\Omega}, x \leq [n/k], j \in \mathcal{S}, \\
& \quad L(z) = i, \text{ and } kx + z(k) < n \\
1 & \text{if } (u, v) = (x, z), x = [n/k], \\
& \quad \text{and } k[n/k] + z(k) = n \\
0 & \text{otherwise},
\end{cases}
$$

$$(4.16)$$

with $d = card(\Omega)$, the size of the state space Ω, equal to

$$\begin{aligned}
d &= 1 + ([n/k] + 1) \times card\{\mathcal{E}\} - l \qquad (4.17) \\
&\quad - card\{([n/k], z) : z \in \mathcal{E}, k[n/k] + z(k) > n\},
\end{aligned}$$

and

(iii) yields the distribution

$$P(X_n(\Lambda) = x) = \boldsymbol{\xi}_0 \boldsymbol{M}^n \boldsymbol{U}'(C_x), \quad x = 0, 1, \cdots, [n/k], \qquad (4.18)$$

where $\boldsymbol{\xi}_0$ is the initial distribution specified by $P(Y_0 = \emptyset) \equiv 1$ and

$$\begin{aligned}
C_{\emptyset} &= [\emptyset], C_0 = [(0, z) : z \in \mathcal{E}] - [(0, \Lambda_i) : i = 1, 2, \cdots, l], \\
C_x &= [(x, z) : z \in \mathcal{E}], \ 1 \leq x < [n/k], \text{ and} \qquad (4.19) \\
C_{[n/k]} &= [([n/k], z) : z \in \mathcal{E}, k[n/k] + z(k) \leq n],
\end{aligned}$$

are the partitions of the state space Ω.

Note that the state space Ω and its size d are functions of n, the structure of the patterns Λ_i, $i = 1, \cdots, l$, and the common pattern length k. The reader may verify that the results of Examples 4.1 to 4.3 follow directly from Theorem 4.1.

Proof.   Given n, since the length of every pattern is k, the maximum number of patterns is $[n/k]$ (under non-overlap counting). The set $\mathcal{E} = \mathcal{S} \cup_{i=1}^l \mathcal{S}(\Lambda_i)$ contains all possible ending blocks generated by $\mathcal{S}$ and all the patterns, and it follows that for forward and backward non-overlap counting, the state space has the form $\{(x, z) : x = 0, 1, \cdots, [n/k], z \in \mathcal{E}\}$. The states $\{(0, \Lambda_i) : i = 1, \cdots, l\}$ are deleted because if the ending block is Λ_i, then there must be at least one pattern Λ_i in the sequence ($x \geq 1$), and so the states $\{(0, \Lambda_i)\}$ are unreachable; by the same token, the states $\{([n/k], z) : k[n/k] + z(k) > n\}$ cannot occur either and can be deleted. Thus the state space Ω of the imbedded Markov chain has the form given by Eq. (4.14), and its size d is determined by Eq. (4.17).

Given $(x, z) \in \Omega$, $0 \leq x \leq [n/k]$, and $kx + z(k) < n$, if $X_t = j \in \mathcal{S}$ and $(u, v) = \; <(x, z), j>_\Omega$, then, as described in Eqs. (4.9) and (4.12), it follows that

$$p_{(x,z)(u,v)} = P(Y_t = (u, v)|Y_{t-1} = (x, z)) = p_{ij},$$

where $i = L(z)$. If $Y_{t-1} = ([n/k], z)$ and $k[n/k] + z(k) = n$, then $t - 1 \equiv n$; for convenience, we assign the transition probabilities for these states to $P(Y_t = ([n/k], z)|Y_{t-1} = ([n/k], z)) \equiv 1$. This completes the construction of Eq. (4.16) and the transition probability matrix $\boldsymbol{M}$. The partitions on the state space Ω, given by Eq. (4.19), are a direct consequence of the definition of the imbedded Markov chain introduced in Eq. (4.3). Therefore, the distribution for the compound pattern Λ in Eq. (4.18) is an immediate consequence of Theorem 2.1. This completes the proof.   □

The above Theorem 4.1 naturally also holds for simple patterns, the special case when $l = 1$. When the pattern lengths k_i, $i = 1, \cdots, l$, are not all equal, the forward and backward principle can still be used to find the distribution for the compound pattern Λ. In principle, the forward and backward counting procedure is applicable for any number of patterns l of varying sizes k_i, but it is not simple to write down the general form of the state space and the transition probability matrix of the imbedded Markov chain. We will discuss this problem in Chapter 5 using the duality relationship between $X_n(\Lambda)$ and the waiting time $W(\Lambda)$.

4.3 Overlap Counting

Let us consider that the sequence $\{X_t\}$ is a homogeneous Markov chain defined on the state space $\mathcal{S} = \{a, b, c\}$ with transition probability matrix

$$\boldsymbol{A} = (p_{ij}), \quad i, j = a, b, c. \tag{4.20}$$

Let Λ be a simple pattern of length k. The basic difference between overlap counting and non-overlap counting is that when the pattern Λ is formed, a part of Λ will be counted toward forming the next pattern Λ under *overlap counting*, up to the last $(k-1)$ trials.

Definition 4.1 An ending block E^o generated by the pattern Λ is the longest ending block $(E^o \neq \Lambda)$ that, after each occurrence of Λ under overlap counting, can be assigned as the initial ending block for the next occurrence of Λ. We write $E^o \cong \Lambda$, with respect to overlap counting.

For example, under overlap counting:

- if $\Lambda = aca$, then $E^o = a$,
- if $\Lambda = abcab$, then $E^o = ab$, and
- if $\Lambda = \underbrace{a \cdots a}_{k}$, then $E^o = \underbrace{a \cdots a}_{k-1}$.

For a pattern such as $\Lambda = abc$, there does not exist an E^o, in which case overlap and non-overlap counting are the same.

Note that under overlap counting, since the first pattern requires k elements and every additional pattern requires only $k - card(E^o)$ elements, the largest possible number of patterns Λ that can occur in n $(n \geq k)$ trials is

$$l_n^o = 1 + \left[\frac{n-k}{k - card(E^o)} \right]. \tag{4.21}$$

In order to illustrate the minor differences arising from the two types of counting procedures, we provide the following example.

Example 4.4 Let us consider the number of patterns $\Lambda = aca$ occurring in $n = 5$ *i.i.d.* three-state trials. Under non-overlap counting, the transition probability matrix $\boldsymbol{M}$ associated with the imbedded Markov chain is

$$
\boldsymbol{M} =
\begin{array}{c}
(0,a)\\(0,b)\\(0,c)\\(0,ab)\\(1,\Lambda)\\(1,a)\\(1,b)\\(1,c)\\(1,ac)
\end{array}
\left(
\begin{array}{ccccccccc}
p_a & p_b & 0 & p_c & 0 & 0 & 0 & 0 & 0\\
p_a & p_b & p_c & 0 & 0 & 0 & 0 & 0 & 0\\
p_a & p_b & p_c & 0 & 0 & 0 & 0 & 0 & 0\\
0 & p_b & p_c & 0 & p_a & 0 & 0 & 0 & 0\\
0 & 0 & 0 & 0 & 0 & p_a & p_b & p_c & 0\\
0 & 0 & 0 & 0 & 0 & p_a & p_b & 0 & p_c\\
0 & 0 & 0 & 0 & 0 & p_a & p_b & p_c & 0\\
0 & 0 & 0 & 0 & 0 & p_a & p_b & p_c & 0\\
0 & 0 & 0 & 0 & 0 & 0 & 0 & 0 & 1
\end{array}
\right),
$$

where $(1,\Lambda) \equiv (1,aca)$. Under overlap counting, the transition probability matrix $\boldsymbol{M}^o$ associated with the imbedded Markov chain is

$$
\boldsymbol{M}^o =
\begin{array}{c}
(0,a)\\(0,b)\\(0,c)\\(0,ab)\\(1,\Lambda)\\(1,a)\\(1,b)\\(1,c)\\(1,ac)\\(2,\Lambda)
\end{array}
\left(
\begin{array}{ccccc|ccccc}
p_a & p_b & 0 & p_c & 0 & 0 & 0 & 0 & 0 & 0\\
p_a & p_b & p_c & 0 & 0 & 0 & 0 & 0 & 0 & 0\\
p_a & p_b & p_c & 0 & 0 & 0 & 0 & 0 & 0 & 0\\
0 & p_b & p_c & 0 & p_a & 0 & 0 & 0 & 0 & 0\\
0 & 0 & 0 & 0 & 0 & p_a & p_b & 0 & p_c & 0\\
0 & 0 & 0 & 0 & 0 & p_a & p_b & 0 & p_c & 0\\
0 & 0 & 0 & 0 & 0 & p_a & p_b & p_c & 0 & 0\\
0 & 0 & 0 & 0 & 0 & p_a & p_b & p_c & 0 & 0\\
0 & 0 & 0 & 0 & 0 & 0 & p_b & p_c & 0 & p_a\\
0 & 0 & 0 & 0 & 0 & 0 & 0 & 0 & 0 & 1
\end{array}
\right).
$$

The primary difference between the two matrices arises after the first pattern has occurred. With probability p_c, the state $(1,\Lambda)$ goes to state $(1,c)$ under non-overlap counting, while $(1,\Lambda)$ goes to $(1,ac)$ under overlap counting, which also entails the additional state $(l_n^o = 2, \Lambda)$. $\diamond$

If $\{X_t\}$ is a homogeneous Markov chain with transition probabilities given by Eq. (4.20), the transition probability matrix $\boldsymbol{M}^o$ of the above example (under overlap counting) becomes

$$
M^o =
\begin{array}{c}
\emptyset \\
(0,a) \\
(0,b) \\
(0,c) \\
(0,ab) \\
(1,\Lambda) \\
(1,a) \\
(1,b) \\
(1,c) \\
(1,ac) \\
(2,\Lambda)
\end{array}
\left(
\begin{array}{cccccc|ccccc}
0 & p_a & p_b & p_c & 0 & 0 & 0 & 0 & 0 & 0 & 0 \\
0 & p_{aa} & p_{ab} & 0 & p_{ac} & 0 & 0 & 0 & 0 & 0 & 0 \\
0 & p_{ba} & p_{bb} & p_{bc} & 0 & 0 & 0 & 0 & 0 & 0 & 0 \\
0 & p_{ca} & p_{cb} & p_{cc} & 0 & 0 & 0 & 0 & 0 & 0 & 0 \\
0 & 0 & p_{bb} & p_{bc} & 0 & p_{ba} & 0 & 0 & 0 & 0 & 0 \\
\hline
0 & 0 & 0 & 0 & 0 & 0 & p_{aa} & p_{ab} & 0 & p_{ac} & 0 \\
0 & 0 & 0 & 0 & 0 & 0 & p_{aa} & p_{ab} & 0 & p_{ac} & 0 \\
0 & 0 & 0 & 0 & 0 & 0 & p_{ba} & p_{bb} & p_{bc} & 0 & 0 \\
0 & 0 & 0 & 0 & 0 & 0 & p_{ca} & p_{cb} & p_{cc} & 0 & 0 \\
0 & 0 & 0 & 0 & 0 & 0 & 0 & p_{cb} & p_{cc} & 0 & p_{ca} \\
0 & 0 & 0 & 0 & 0 & 0 & 0 & 0 & 0 & 0 & 1
\end{array}
\right),
$$

where p_a, p_b, and p_c are the given transition probabilities from state $\emptyset$ to states $(0,a)$, $(0,b)$, and $(0,c)$, respectively. Again, the extension from $i.i.d.$ to Markov-dependent trials remains straightforward.

The concept considered in the above example can be extended to the case of overlap up to the last d trials $(1 \leq d \leq k-1)$, as introduced by Aki and Hirano (2000). It is a significant advantage of the finite Markov chain imbedding technique that the extension from non-overlap counting to overlap counting is direct and simple.

4.4 Series Pattern

The distribution of the number of series patterns $\Lambda = \Lambda_1 * \Lambda_2$ can be obtained in almost the same manner as for a simple pattern, with minor modifications at the state after the first pattern Λ_1 has occurred.

Example 4.5 Let us consider a sequence of $n = 5$ $i.i.d.$ three-state trials drawn from $\mathcal{S} = \{a,b,c\}$, and the series pattern $\Lambda = ab * cc$ generated by the two simple patterns $\Lambda_1 = ab$ and $\Lambda_2 = cc$. Define the ending block set $\mathcal{E} = \{a, \bar{a}, ab*, ab*c, ab*cc\}$ and the state space

$$\Omega = \{\emptyset, (0,a), (0,\bar{a}), (0,ab*), (0,ab*c), (1,ab*cc), (1,a), (1,\bar{a})\},$$

where $\bar{a}$ stands for b or c, and $ab*$ represents $ab*a$ or $ab*b$. The corresponding imbedded Markov chain $\{Y_t\}$ has the transition probability matrix M

given by

$$
M = \begin{array}{c}
\emptyset \\
(0,a) \\
(0,\bar{a}) \\
(0,ab*) \\
(0,ab*c) \\
(1,ab*cc) \\
(1,a) \\
(1,\bar{a})
\end{array}
\begin{pmatrix}
0 & p_a & p_b+p_c & 0 & 0 & 0 & 0 & 0 \\
0 & p_a & p_c & p_b & 0 & 0 & 0 & 0 \\
0 & p_a & p_b+p_c & 0 & 0 & 0 & 0 & 0 \\
0 & 0 & 0 & p_a+p_b & p_c & 0 & 0 & 0 \\
0 & 0 & 0 & p_a+p_b & 0 & p_c & 0 & 0 \\
0 & 0 & 0 & 0 & 0 & 0 & p_a & p_b+p_c \\
0 & 0 & 0 & 0 & 0 & 0 & 1 & 0 \\
0 & 0 & 0 & 0 & 0 & 0 & 0 & 1
\end{pmatrix}.
$$

Hence

$$
P(X_n(\Lambda) = x) = \boldsymbol{\xi}_0 \boldsymbol{M}^5 \boldsymbol{U}'(C_x), \quad x = 0, 1.
$$

Note that if Y_t is at the state $(0, ab*)$ or $(0, ab * c)$, this means that the pattern $\Lambda_1 = ab$ has occurred before or at the t-th trial. Now if the realization of X_{t+1} is a or b, then Y_{t+1} has to be at state $(0, ab*)$, an event occurring with transition probability $p_a + p_b$, and if the realization of X_{t+1} is c, then Y_{t+1} advances to state $(0, ab*c)$ or $(1, ab*cc)$, respectively, occurring with transition probability p_c. ◇

The above example highlights the differences between series and simple patterns with regard to their transition probability matrices of the imbedded Markov chains. By slightly enlarging the state space Ω, replacing $(i, \bar{a})$, $i = 0, 1$, with (i, b) and (i, c), and replacing $(0, ab*)$ with $(0, ab * a)$ and $(0, ab * b)$, the above example can be easily extended to the case of Markov-dependent three-state trials.

4.5   Joint Distribution

Finding the joint distribution of two numbers of runs, say $X_n(\Lambda_1)$ and $X_n(\Lambda_2)$, in a sequence of two- or multi-state trials $\{X_t\}$ using the finite Markov chain imbedding technique is similar to finding the exact distribution of $X_n(\Lambda)$ introduced in Section 4.2. In general, the imbedded Markov chain $\{Y_t\}$ associated with the joint distribution of $X_n(\Lambda_1)$ and $X_n(\Lambda_2)$ has the form

$$
Y_t = (X_t(\Lambda_1), X_t(\Lambda_2), E_t), \quad t = 1, 2, \cdots, n. \tag{4.22}
$$

The state space Ω and the ending block E_t for $\{Y_t\}$ depend very much on the structure of the patterns Λ_1 and Λ_2. The transition probability matrices $\boldsymbol{M}_t$ of the imbedded Markov chain can be constructed using the same principles described in earlier sections. In the following, we give one example to demonstrate the procedure for finding the joint distribution.

Example 4.6 Let X_n be the total number of success runs $X_n(S)$ and failure runs $X_n(F)$ in a sequence of n two-state trials. For every n, $X_n = X_n(S) + X_n(F)$, and the numbers of runs $X_n(S)$ and $X_n(F)$ are related in the following way: if there are x success runs, then there can only be $x + 1$, x, or $x - 1$ failure runs. It follows that there can only be four types of states $Y_t = (X_t(S), X_t(F), E_t)$, where the ending block E_t is either S or F: (i) $(x, x - 1, S)$, (ii) $(x, x + 1, F)$, (iii) (x, x, S), and (iv) (x, x, F).

Consider the outcomes of ten two-state trials $\omega = (SSFFSFSSSF)$. The realization of the imbedded Markov chain $\{Y_t\}$ is $\{Y_1 = (1, 0, S)$, $Y_2 = (1, 0, S)$, $Y_3 = (1, 1, F)$, $Y_4 = (1, 1, F)$, $Y_5 = (2, 1, S)$, $Y_6 = (2, 2, F)$, $Y_7 = (3, 2, S)$, $Y_8 = (3, 2, S)$, $Y_9 = (3, 2, S)$, $Y_{10} = (3, 3, F)\}$. The state space Ω has the form $\Omega = \{(1, 0, S), (0, 1, F), (1, 1, S), (1, 1, F), \cdots, (l_n, l_n, S), (l_n, l_n, F)\}$, where $l_n = [(n + 1)/2]$. For the case of independent but non-identically distributed two-state trials, the definition of Y_t yields the transition probability matrices, for $t = 2, 3, \cdots, n$,

$$
\boldsymbol{M}_t =
\begin{array}{c}
(1, 0, S) \\
(0, 1, F) \\
(1, 1, S) \\
(1, 1, F) \\
\cdot \\
\cdot \\
\cdot \\
(l_n, l_n - 1, S) \\
(l_n - 1, l_n, F) \\
(l_n, l_n, S) \\
(l_n, l_n, F)
\end{array}
\left(
\begin{array}{ccccccccc}
p_t & 0 & 0 & q_t & & & & & \\
q_t & p_t & 0 & 0 & & & \boldsymbol{O} & & \\
 & p_t & 0 & 0 & q_t & & & & \\
 & q_t & p_t & 0 & & & & & \\
 & & & & \cdot & \cdot & \cdot & \cdot & \\
 & & & & & \cdot & \cdot & \cdot & \cdot \\
 & & & & & & \cdot & \cdot & \cdot \\
 & & & & & p_t & 0 & 0 & q_t \\
 & & & & & q_t & p_t & 0 & \\
 & & \boldsymbol{O} & & & & 1 & 0 & \\
 & & & & & & & & 1
\end{array}
\right)
$$

$$(4.23)$$

Given $\boldsymbol{\xi}_1 = (p_1, q_1, 0, \cdots, 0)$, it follows that the joint distribution of $X_n(S)$ and $X_n(F)$ is given by

$$
P(X_n(S) = x, X_n(F) = y | \boldsymbol{\xi}_1) = \boldsymbol{\xi}_1 (\prod_{t=2}^{n} \boldsymbol{M}_t) \boldsymbol{U}^{'}(C_{(x,y)}), \tag{4.24}
$$

where, if $y = x + 1$, then $C_{(x,x+1)} = \{(x, x + 1, F)\}$, if $y = x - 1$, then $C_{(x,x-1)} = \{(x, x - 1, S)\}$, if $y = x$, then $C_{(x,x)} = \{(x, x, S), (x, x, F)\}$, and $C_{(x,y)} = \{\emptyset\}$ elsewhere. $\diamond$

Again, with some simple modifications to the transition probability matrices, the above results also hold for *i.i.d.* as well as homogeneous and non-homogeneous Markov-dependent two-state trials. The marginal distributions of $X_n(S)$ and $X_n(F)$ can be obtained by projecting the joint distribution onto the partitions generated by the random variables $X_n(S)$ and $X_n(F)$, respectively. More generally, the joint distribution of $l > 2$ random variables $X_n(\Lambda_1), X_n(\Lambda_2), \cdots, X_n(\Lambda_l)$ can be obtained in the same fashion with an $(l + 1)$-dimensional Markov chain $\{Y_t = (X_t(\Lambda_1), \cdots, X_t(\Lambda_l), E_t)\}$.

Chapter 5

Waiting-Time Distributions

5.1 Introduction

The geometric distribution is the simplest waiting-time distribution of a success in a sequence of Bernoulli trials. Its probability distribution function is given by

$$P(W(S) = n) = pq^{n-1}, \quad n = 1, 2, \cdots. \tag{5.1}$$

The waiting time of the r-th success has a negative binomial distribution given by

$$P(W(r, S) = n) = \binom{n-1}{r-1} p^r q^{n-r}, \quad n = r, r+1, \cdots.$$

For the pattern $\Lambda = S \cdots S$ of k consecutive successes, as discussed in Section 3.7, the waiting time $W(\Lambda)$ has a *geometric distribution of order k*. This distribution has been studied by many authors, including, for example, Aki (1985), Aki and Hirano (1988), Aki, Kuboki and Hirano (1984), Feller (1968), Hirano (1986), Hirano and Aki (1987), Koutras (1996b, 1997a), Mohanty (1994), Philippou (1986), and Philippou, Georghiou and Philippou (1983). In view of the form of the geometric distribution of order k given in Section 3.7, we define a large family of geometric-type distributions.

Definition 5.1 A waiting-time random variable $W(\Lambda)$ has a *general geometric distribution* if $W(\Lambda)$ is Markov chain imbeddable with probability distribution function

$$P(W(\Lambda) = n) = \boldsymbol{\xi} \boldsymbol{N}^{n-1} (\boldsymbol{I} - \boldsymbol{N}) \boldsymbol{1}', \quad n = 1, 2, \cdots, \tag{5.2}$$

where N is the essential transition probability submatrix (defined in Section 2.6) of M corresponding to the imbedded Markov chain $\{Y_t\}$.

For example, the waiting-time random variable $W(r, \Lambda)$, the waiting time for the r-th occurrence of a compound pattern $\Lambda = \cup_{i=1}^{l} \Lambda_i$ in homogeneous Markov-dependent multi-state trials, also has a general geometric distribution, *i.e.* its distribution can be computed via Eq. (5.2). The class of general geometric distributions is extremely large. It covers all well-known discrete waiting-time distributions, including the waiting-time distributions of the extended Stirling family (see Nishimura and Sibuya 1997). Table 5.1 lists some general geometric distributions commonly seen in practice for Bernoulli trials:

Table 5.1   Some common general geometric waiting-time distributions for Bernoulli trials.

Distribution	Compound Pattern	R.V.
Geometric, $G(p)$	$\Lambda = S$	$W(\Lambda)$
Geometric of order k, $G_k(p)$	$\Lambda = S \cdots S$ of length k	$W(\Lambda)$
Negative Binomial, $NB(r, p)$	r-th occurrence of $\Lambda = S$	$W(r, \Lambda)$
Negative Binomial		$W(r, \Lambda)$
of order k, $NB_k(r, p)$	r-th occurences of $\Lambda = S \cdots S$	
Sooner	$\Lambda = \cup_{i=1}^{l} \Lambda_i$, $l \geq 2$	$W(\Lambda)$
Scan	(defined in Section 5.9)	$S_n(r)$

It appears that all well-known waiting-time distributions can be viewed as a special case of the general geometric distribution for the r-th occurrence of a compound pattern $(r \geq 1)$. In this chapter, we will focus on constructing the imbedded Markov chain $\{Y_t\}$ for the waiting time of a compound pattern. The average waiting times and the generating functions for various waiting-time distributions will be obtained through Eq. (5.2). Further, we also study the large deviation approximation for the tail probability $P(W(\Lambda) \geq n)$.

The dual relationship between the random variables $X_n(\Lambda)$ and $W(r, \Lambda)$,

$$\{X_n(\Lambda) < r\} \text{ if and only if } \{W(r, \Lambda) > n\},$$

gives the probability identity

$$P(X_n(\Lambda) < r) = P(W(r, \Lambda) > n). \tag{5.3}$$

This identity will provide a way to compute the exact distribution of the number of occurrences of pattern Λ in n trials by computing the tail probabilities of its corresponding waiting-time random variable. For example, the distributions of the longest run statistic and the scan statistic are obtained in this book through such dual relationships.

5.2 The Waiting Time of A Simple Pattern

Let $W(\Lambda)$ be the waiting time for the first occurrence of a simple pattern $\Lambda = b_{i_1} b_{i_2} \cdots b_{i_k}$, with $i_j \in \{1, \cdots, m\}$, in a sequence of *i.i.d.* m-state trials $\{X_t\}$. The imbedded Markov chain $\{Y_t\}$ associated with the waiting-time random variable $W(\Lambda)$ is a homogeneous Markov chain defined on the state space

$$\Omega = \{\emptyset\} \cup \mathbb{S} \cup \mathbb{S}(\Lambda) = \{\emptyset, b_1, b_2, \cdots, b_m, b_{i_1} b_{i_2}, \cdots, b_{i_1} b_{i_2} \cdots b_{i_{k-1}}, \alpha\}$$

$(\alpha = \Lambda)$, with a transition probability matrix of the form

$$\boldsymbol{M} = (p_{u,v}) = \begin{matrix} \Omega - \alpha \\ \alpha \end{matrix} \left(\begin{array}{c|c} \boldsymbol{N} & \boldsymbol{C} \\ \hline 0 & 1 \end{array} \right), \tag{5.4}$$

where α is an absorbing state and $Y_t = \alpha$ means that the pattern Λ has occurred on or before the t-th trial. For convenience, a dummy state "$\emptyset$" is added in the state space Ω so that we can assume $P(Y_0 = \emptyset) = 1$ as the initial distribution for the imbedded Markov chain. We will provide the details of how to construct the transition probabilities $p_{u,v}$ in the next section.

Theorem 5.1 *If $\{X_t\}$ is a sequence of i.i.d. m-state trials, then the distribution of the waiting time for a simple pattern $\Lambda = b_{i_1} b_{i_2} \cdots b_{i_k}$ is given by*

$$P(W(\Lambda) = n) = \boldsymbol{\xi} \boldsymbol{N}^{n-1} \boldsymbol{C} = \boldsymbol{\xi} \boldsymbol{N}^{n-1} (\boldsymbol{I} - \boldsymbol{N}) \boldsymbol{1}', \tag{5.5}$$

where the matrix $\boldsymbol{N}$ and the column vector $\boldsymbol{C}$ are defined by Eq. (5.4).

Proof.   Given $Y_0 = \emptyset$, since $\{Y_t\}$ is the imbedded Markov chain for $W(\Lambda)$, it follows that $W(\Lambda) = n$ if and only if the Markov chain $\{Y_t\}$ enters the state α for the first time at the n-th trial; *i.e.*

$$P(W(\Lambda) = n | Y_0 = \emptyset) = P(Y_n = \alpha, Y_{n-1} \neq \alpha, \cdots, Y_1 \neq \alpha | Y_0 = \emptyset). \quad (5.6)$$

The result (5.5) follows immediately from Theorem 2.2, Eqs. (5.4) and (5.6), and the identity $(\boldsymbol{I} - \boldsymbol{N})\boldsymbol{1}' = \boldsymbol{C}$ (rows summing to one).   □

Example 5.1   Consider a sequence of *i.i.d.* four-state trials $\{X_t\}$ with possible outcomes a, c, g, and t, occurring with probabilities p_a, p_c, p_g, and p_t, respectively. Suppose that we are interested in finding the waiting-time distribution for the pattern $\Lambda = acac$. The distribution of the waiting-time random variable $W(\Lambda)$ is determined by Eq. (5.5) through the imbedded Markov chain $\{Y_t\}$ defined on the state space $\Omega = \{\emptyset\} \cup \mathcal{S} \cup \mathcal{S}(\Lambda) = \{\emptyset, a, c, g, t, ac, aca, \alpha\}$ with transition probability matrix

$$\boldsymbol{M} = \begin{array}{c} \emptyset \\ a \\ c \\ g \\ t \\ ac \\ aca \\ \alpha \end{array} \left(\begin{array}{ccccccc:c} 0 & p_a & p_c & p_g & p_t & 0 & 0 & 0 \\ 0 & p_a & 0 & p_g & p_t & p_c & 0 & 0 \\ 0 & p_a & p_c & p_g & p_t & 0 & 0 & 0 \\ 0 & p_a & p_c & p_g & p_t & 0 & 0 & 0 \\ 0 & p_a & p_c & p_g & p_t & 0 & 0 & 0 \\ 0 & 0 & p_c & p_g & p_t & 0 & p_a & 0 \\ \hdashline 0 & p_a & 0 & p_g & p_t & 0 & 0 & p_c \\ 0 & 0 & 0 & 0 & 0 & 0 & 0 & 1 \end{array} \right).$$

If $\{X_t\}$ is Markov dependent, the probability distribution of $W(\Lambda)$ can also be obtained through Eq. (5.5), with some simple modifications to the transition probabilities in $\boldsymbol{M}$ (without changing the state space). The explicit form of $\boldsymbol{M}$ for Markov-dependent trials as well as compound patterns will be given in the next section.   ◇

5.3   The Waiting Time of A Compound Pattern

Before extending Theorem 5.1 to the more general case where $\{X_t\}$ is an m-state homogeneous Markov chain defined on $\mathcal{S} = \{b_1, \cdots, b_m\}$ with transition probabilities (p_{ij}) and where Λ is a compound pattern of l distinct simple patterns ($\Lambda = \cup_{i=1}^{l} \Lambda_i$), we first discuss a specific extension of Example 5.1 to illustrate the imbedding procedure. Let $\Lambda = \Lambda_1 \cup \Lambda_2$ be a compound pattern composed of two distinct simple patterns $\Lambda_1 = acac$ and $\Lambda_2 = ctc$. Define the imbedded homogeneous Markov chain $\{Y_t\}$ on the state space

$$\Omega = \{\emptyset\} \cup \mathcal{S} \cup \mathcal{S}(\Lambda_1) \cup \mathcal{S}(\Lambda_2) = \{\emptyset, a, c, g, t, ac, aca, ct, \alpha_1, \alpha_2\},$$

where α_1 and α_2 are absorbing states corresponding to Λ_1 and Λ_2, respectively. Given $Y_{t-1} = u \in \Omega$, say $u = ac$, it follows from the forward and backward principle (no pattern Λ has occurred in counting forward and the longest ending block in the state space Ω is obtained from counting backward) that there are only four possible outcomes of $Y_t = v$ corresponding to the outcomes $X_t = a$, c, g, and t:

$$
\begin{array}{cccccl}
& \text{time } t-1 && \text{time } t && \\
& u && v && \\
&& \xrightarrow{p_{ca}} & aca & \text{if } X_t = a & \\
& ac & \xrightarrow{p_{cc}} & c & \text{if } X_t = c & \\
&& \xrightarrow{p_{cg}} & g & \text{if } X_t = g & \\
&& \xrightarrow{p_{ct}} & ct & \text{if } X_t = t, &
\end{array}
\tag{5.7}
$$

where p_{ij}, $i, j = a, c, g, t$, are the transition probabilities of the Markov chain $\{X_t\}$. Thus the transition probabilities of $\{Y_t\}$ are

$$P(Y_t = v | Y_{t-1} = ac) = \begin{cases} p_{ca} & \text{if } X_t = a, v = aca \\ p_{cc} & \text{if } X_t = c, v = c \\ p_{cg} & \text{if } X_t = g, v = g \\ p_{ct} & \text{if } X_t = t, v = ct. \end{cases} \tag{5.8}$$

In view of Eqs. (5.7) and (5.8), the imbedded Markov chain $\{Y_t\}$ has the transition probability matrix

$$
M = \begin{array}{c} \emptyset \\ a \\ c \\ g \\ t \\ ac \\ ct \\ aca \\ \alpha_1 \\ \alpha_2 \end{array}
\left(
\begin{array}{cccccccc:cc}
0 & p_a & p_c & p_g & p_t & 0 & 0 & 0 & 0 & 0 \\
0 & p_{aa} & 0 & p_{ag} & p_{at} & p_{ac} & 0 & 0 & 0 & 0 \\
0 & p_{ca} & p_{cc} & p_{cg} & 0 & 0 & p_{ct} & 0 & 0 & 0 \\
0 & p_{ga} & p_{gc} & p_{gg} & p_{gt} & 0 & 0 & 0 & 0 & 0 \\
0 & p_{ta} & p_{tc} & p_{tg} & p_{tt} & 0 & 0 & 0 & 0 & 0 \\
0 & 0 & p_{cc} & p_{cg} & 0 & 0 & p_{ct} & p_{ca} & 0 & 0 \\
0 & p_{ta} & 0 & p_{tg} & p_{tt} & 0 & 0 & 0 & 0 & p_{tc} \\
0 & p_{aa} & 0 & p_{ag} & p_{at} & 0 & 0 & 0 & p_{ac} & 0 \\
\hdashline
0 & 0 & 0 & 0 & 0 & 0 & 0 & 0 & 1 & 0 \\
0 & 0 & 0 & 0 & 0 & 0 & 0 & 0 & 0 & 1
\end{array}
\right)
$$

$$= \quad \begin{matrix} \Omega - A \\ A \end{matrix} \left(\begin{array}{c|c} N & C \\ \hline 0 & I \end{array} \right), \tag{5.9}$$

where A denotes the collection of absorbing states α_1 and α_2.

For the general case, we define the state space

$$\Omega = \{\emptyset\} \cup \mathcal{S} \cup_{i=1}^{l} \mathcal{S}(\Lambda_i), \tag{5.10}$$

where $\mathcal{S} = \{b_1, \cdots, b_m\}$ and $\mathcal{S}(\Lambda_i) = \{$all sequential subpatterns of $\Lambda_i\}$. Let α_i, $i = 1, \cdots, l$, represent the absorbing states corresponding to the simple patterns $\Lambda_1, \cdots, \Lambda_l$. Note that the state space Ω discussed earlier in this section is a special case of Eq. (5.10) with $\Lambda_1 = acac$ and $\Lambda_2 = ctc$. Let $A = \{\alpha_1, \cdots, \alpha_l\}$ be the set of all absorbing states. To generalize Eq. (5.8), for given $Y_{t-1} = u \in \Omega - A - \{\emptyset\}$ and $X_t = z \in \mathcal{S}$, we define

$$
\begin{aligned}
v &= \; <u, z>_{\Omega} \\
&= \; \text{the longest ending block in } \Omega \text{ with respect to the} \\
&\quad \text{forward and backward counting procedure.} \tag{5.11}
\end{aligned}
$$

Further, for each $u \in \Omega - A - \{\emptyset\}$, we define a subset

$$[u : \mathcal{S}] = \{v : v \in \Omega, v = <u, z>_{\Omega}, z \in \mathcal{S}\}. \tag{5.12}$$

Theorem 5.2 *If $\{X_t\}$ is a sequence of homogeneous Markov-dependent m-state trials and $\Lambda = \cup_{i=1}^{l} \Lambda_i$ is a compound pattern consisting of l simple patterns $\Lambda_1, \cdots, \Lambda_l$, then*

(i) *the imbedded Markov chain $\{Y_t\}$ defined on the state space Ω given by Eq. (5.10) has a transition probability matrix*

$$M = (p_{u,v})_{d \times d} = \left(\begin{array}{c|c} N_{(d-l) \times (d-l)} & C \\ \hline 0 & I \end{array} \right)_{d \times d}, \tag{5.13}$$

with $u, v \in \Omega$ and transition probabilities $p_{u,v}$ given by

$$p_{u,v} = P(Y_t = v | Y_{t-1} = u) = \begin{cases} p_z & \text{if } u = \emptyset, v = z, z \in \mathcal{S} \\ p_{xz} & \text{if } u \in \Omega - A - \{\emptyset\}, \\ & \quad v \in [u : \mathcal{S}] \text{ and } X_t = z \\ 1 & \text{if } u \in A \text{ and } v = u \\ 0 & \text{otherwise,} \end{cases} \tag{5.14}$$

where x stands for the last symbol of u, p_z are the initial probabilities from state $\emptyset$ to z, p_{xz} are the transition probabilities of the Markov chain $\{X_t\}$, and $d = card(\Omega)$,

(ii) given the initial distribution $\boldsymbol{\xi}_0 = (\boldsymbol{\xi} : \mathbf{0})$ with $P(Y_0 = \emptyset) = 1$, the waiting-time distribution of the compound pattern Λ is given by

$$P(W(\Lambda) = n) = \boldsymbol{\xi} \boldsymbol{N}^{n-1}(\boldsymbol{I} - \boldsymbol{N})\mathbf{1}', \qquad (5.15)$$

where $\boldsymbol{N}$ is defined by Eqs. (5.13) and (5.14), and
(iii) for $i = 1, \cdots, l$,

$$P(W(\Lambda) = n, W(\Lambda_i) = n) = \boldsymbol{\xi} \boldsymbol{N}^{n-1} \boldsymbol{C}_i, \qquad (5.16)$$

where $\boldsymbol{C}_i$ is the i-th column vector of the matrix $\boldsymbol{C}$.

For example, taking $u = ac$, Eq. (5.14) leads to Eq. (5.8) of our example at the beginning of this section. Further taking every $u \in \Omega$, Eq. (5.14) yields the transition probability matrix $\boldsymbol{M}$ given by Eq. (5.9).

Equations (5.10) and (5.14) provide a simple way to construct the imbedded Markov chain $\{Y_t\}$ on the state space Ω with transition probability matrix $\boldsymbol{M}$. Given $\Lambda = \cup_{i=1}^{l}\Lambda_i$, an automatic computer algorithm to construct the state space Ω and the transition probability matrix $\boldsymbol{M}$, as well as to compute the distribution of the waiting time $W(\Lambda)$ and its generating function, is given by Fu and Chang (2002). The algorithm is based on Eqs. (5.10)-(5.16).

Note that if $\{X_t\}$ is a sequence of $i.i.d.$ m-state trials, then Eq. (5.14) reduces to

$$p_{u,v} = P(Y_t = v | Y_{t-1} = u) = \begin{cases} p_z & \text{if } X_t = z,\ v \in [u:\mathcal{S}] \text{ and } u \in \Omega - A \\ 1 & \text{if } u \in A \text{ and } v = u \\ 0 & \text{otherwise.} \end{cases}$$

$$(5.17)$$

Proof. In view of the concept introduced in Eqs. (5.7) and (5.8), the results given by Eqs. (5.13) and (5.14) are direct consequences of the structure of the state space Ω, the forward and backward counting procedure, and the fact that $\{X_t\}$ is a homogeneous Markov chain. Since, for every $n = 1, 2, \cdots$,

$$Y_n \in \Omega - A \iff W(\Lambda) > n, \qquad (5.18)$$

it follows that

$$
\begin{aligned}
P(W(\Lambda) > n) &= P(Y_n \in \Omega - A) \\
&= \boldsymbol{\xi}_0 \boldsymbol{M}^n \boldsymbol{U}'(\Omega - A), \qquad (5.19)
\end{aligned}
$$

where $\boldsymbol{\xi}_0 = (\boldsymbol{\xi} : \boldsymbol{0})$ and $\boldsymbol{U}(\Omega - A) = (1, \cdots, 1, 0, \cdots, 0) = (\boldsymbol{1} : \boldsymbol{0})$. Hence our result (ii) can be concluded from Theorem 2.2 and Eq. (5.19); *i.e.*

$$
\begin{aligned}
P(W(\Lambda) = n) &= P(W(\Lambda) > n - 1) - P(W(\Lambda) > n) \\
&= \boldsymbol{\xi}_0 \boldsymbol{M}^{m-1} \boldsymbol{U}'(\Omega - A) - \boldsymbol{\xi}_0 \boldsymbol{M}^n \boldsymbol{U}'(\Omega - A) \\
&= \boldsymbol{\xi} \boldsymbol{N}^{n-1} \boldsymbol{1}' - \boldsymbol{\xi} \boldsymbol{N}^n \boldsymbol{1}' \\
&= \boldsymbol{\xi} \boldsymbol{N}^{n-1}(\boldsymbol{I} - \boldsymbol{N}) \boldsymbol{1}'.
\end{aligned}
$$

From the definition of $W(\Lambda)$ and $\{Y_t\}$, the following two events are equivalent:

$$
\{W(\Lambda) = n, W(\Lambda_i) = n\} \iff \{Y_t \in \Omega - A, t = 1, \cdots, n-1, \text{ and } Y_n = \alpha_i\}. \tag{5.20}
$$

Hence, it follows from Eq. (5.20) and Theorem 2.4 that

$$
\begin{aligned}
P(W(\Lambda) = n, W(\Lambda_i) = n) &= P(Y_n = \alpha_i, Y_t \in \Omega - A, t = 1, \cdots, n-1) \\
&= \sum_{u \in \Omega - A} p_{u,\alpha_i} P(Y_{n-1} = u) \\
&= \boldsymbol{\xi} \boldsymbol{N}^{n-1} \boldsymbol{C}_i.
\end{aligned}
$$

This completes the proof.   □

Note that

$$
\begin{aligned}
P(W(\Lambda) = n) &= \sum_{i=1}^{l} P(W(\Lambda) = n, W(\Lambda_i) = n) \\
&= \sum_{i=1}^{l} \boldsymbol{\xi} \boldsymbol{N}^{n-1} \boldsymbol{C}_i = \boldsymbol{\xi} \boldsymbol{N}^{n-1}(\boldsymbol{I} - \boldsymbol{N}) \boldsymbol{1}'.
\end{aligned}
$$

Although the above theorem is proved only under the assumption that $\{X_t\}$ is a homogeneous Markov chain, the theorem remains true after minor modifications if the trials X_t are: (i) *i.i.d.* (Theorem 5.1 is a special case for a simple pattern), (ii) independent but non-identically distributed, or (iii) non-homogeneous Markov-dependent. We provide the following example to illustrate this point.

Example 5.2 (A) Let $\{X_t\}$ be a sequence of Bernoulli trials and Λ a compound pattern consisting of $\Lambda_1 = SSS$ and $\Lambda_2 = FFF$. The imbedded Markov chain $\{Y_t\}$ associated with the waiting-time random variable $W(\Lambda)$ is defined on the state space $\Omega = \{\emptyset, S, SS, F, FF, \alpha_1, \alpha_2\}$ with transition probability matrix

$$
M = \begin{array}{c}
\emptyset \\ S \\ F \\ SS \\ FF \\ \alpha_1 \\ \alpha_2
\end{array}
\left(
\begin{array}{ccccc|cc}
0 & p_S & p_F & 0 & 0 & 0 & 0 \\
0 & 0 & p_F & p_S & 0 & 0 & 0 \\
0 & p_S & 0 & 0 & p_F & 0 & 0 \\
0 & 0 & p_F & 0 & 0 & p_S & 0 \\
0 & p_S & 0 & 0 & 0 & 0 & p_F \\
\hline
0 & 0 & 0 & 0 & 0 & 1 & 0 \\
0 & 0 & 0 & 0 & 0 & 0 & 1
\end{array}
\right)
= \left(
\begin{array}{c|c}
N & C \\
\hline
0 & I
\end{array}
\right).
$$

The distribution of the waiting time $W(\Lambda)$ can be obtained from Eq. (5.15).

(B) If $\{X_t\}$ is a sequence of non-homogeneous Markov-dependent trials, then the transition probability matrices of the imbedded Markov chain $\{Y_t\}$ for the waiting-time random variable $W(\Lambda)$ are, for $t = 1, 2, \cdots, n$,

$$
M_t = \begin{array}{c}
\emptyset \\ S \\ F \\ SS \\ FF \\ \alpha_1 \\ \alpha_2
\end{array}
\left(
\begin{array}{ccccc|cc}
0 & p_S & p_F & 0 & 0 & 0 & 0 \\
0 & 0 & p_{SF}(t) & p_{SS}(t) & 0 & 0 & 0 \\
0 & p_{FS}(t) & 0 & 0 & p_{FF}(t) & 0 & 0 \\
0 & 0 & p_{SF}(t) & 0 & 0 & p_{SS}(t) & 0 \\
0 & p_{FS}(t) & 0 & 0 & 0 & 0 & p_{FF}(t) \\
\hline
0 & 0 & 0 & 0 & 0 & 1 & 0 \\
0 & 0 & 0 & 0 & 0 & 0 & 1
\end{array}
\right)
$$

$$
= \left(
\begin{array}{c|c}
N(t) & C(t) \\
\hline
0 & I
\end{array}
\right). \tag{5.21}
$$

The distribution of $W(\Lambda)$ is given by

$$
P(W(\Lambda) = n) = \xi \left(\prod_{t=1}^{n-1} N(t) \right) (I - N(n)) \, 1', \tag{5.22}
$$

and, for $i = 1, 2$,

$$
P(W(\Lambda) = n \text{ and } W(\Lambda_i) = n) = \xi \left(\prod_{t=1}^{n-1} N(t) \right) C_i(n), \tag{5.23}
$$

where the matrices $N(t)$ and $C(t)$ are defined by Eq. (5.21), and $C_i(n)$, $i = 1, 2$, are the columns of matrix $C(n)$.   ◇

Note that, for a given k $(1 \leq k \leq n)$, the probability of the longest run (success or failure run) $L_n < k$ is equal to the probability of the waiting time $W(\Lambda) > n$ with $\Lambda = \Lambda_1 \cup \Lambda_2$, $\Lambda_1 = \underbrace{F \cdots F}_{k}$ and $\Lambda_2 = \underbrace{S \cdots S}_{k}$, so that

$$P(L_n < k) = \boldsymbol{\xi} \left(\prod_{t=1}^{n} N(t) \right) \mathbf{1}'.$$

Our Example 5.2 is only a special case for two-state trials. Similarly, the lower-tail probability of the longest-run distribution in multi-state trials can also be cast in terms of the upper-tail probability of the waiting-time distribution for a compound pattern.

Since $\Lambda = \cup_{i=1}^{l} \Lambda_i$, it follows from the definition of the compound pattern that the waiting time $W(\Lambda)$ is related to the individual waiting times $W(\Lambda_i)$ in the following way:

$$W(\Lambda) = \min(W(\Lambda_1), \cdots, W(\Lambda_l)).$$

This is the reason why $W(\Lambda)$ is referred to as the *sooner waiting time* in the literature. It has been studied by, for example, Balasubramanian, Viveros and Balakrishnan (1993), Aki (1997), Aki, Balakrishnan, and Mohanty (1996), Aki and Hirano (2000), Ebneshahrashoob and Sobel (1990), Han and Aki (2000a), Koutras (1997b), Koutras and Alexandrou (1997b), Ling (1992), Ling and Low (1993), and Uchida and Aki (1995).

5.4   Probability Generating Function

In Chapter 3, the probability generating function of the waiting-time random variable $W(\Lambda)$ for k consecutive successes $\Lambda = S \cdots S$ in Bernoulli trials,

$$\varphi_W(s) = \frac{p^k s^k (1 - ps)}{1 - s + q p^k s^{k+1}},$$

was developed through the recursive equation (3.31). For *i.i.d.* and homogeneous Markov-dependent multi-state trials, the probability generating function and the mean waiting time for a simple pattern Λ are given by the following theorem.

Theorem 5.3 *Suppose that M is the transition probability matrix, in the form of Eq. (5.4), of the imbedded Markov chain $\{Y_t\}$ for the waiting-time random variable $W(\Lambda)$ of a simple pattern Λ, then*

(i)

$$\varphi_W(s) = 1 + (s-1)\boldsymbol{\xi}(\boldsymbol{I} - s\boldsymbol{N})^{-1}\boldsymbol{1}', \qquad (5.24)$$

and

(ii)

$$EW(\Lambda) = \boldsymbol{\xi}(\boldsymbol{I} - \boldsymbol{N})^{-1}\boldsymbol{1}'. \qquad (5.25)$$

Proof. It follows from the definition of the generating function and from Theorem 5.1 that

$$
\begin{aligned}
\varphi_W(s) &= \sum_{n=1}^{\infty} s^n P(W(\Lambda) = n) \\
&= \sum_{n=1}^{\infty} s^n [P(W(\Lambda) > n-1) - P(W(\Lambda) > n)] \\
&= \sum_{n=1}^{\infty} s^n \boldsymbol{\xi} \boldsymbol{N}^{n-1} \boldsymbol{1}' - \sum_{n=1}^{\infty} s^n \boldsymbol{\xi} \boldsymbol{N}^n \boldsymbol{1}' \\
&= s \sum_{n=1}^{\infty} s^{n-1} \boldsymbol{\xi} \boldsymbol{N}^{n-1} \boldsymbol{1}' - \sum_{n=1}^{\infty} s^n \boldsymbol{\xi} \boldsymbol{N}^n \boldsymbol{1}' \\
&= s \sum_{n=0}^{\infty} s^n \boldsymbol{\xi} \boldsymbol{N}^n \boldsymbol{1}' - \sum_{n=0}^{\infty} s^n \boldsymbol{\xi} \boldsymbol{N}^n \boldsymbol{1}' + 1 \\
&= 1 + (s-1)\boldsymbol{\xi} \left(\sum_{n=0}^{\infty} s^n \boldsymbol{N}^n \right) \boldsymbol{1}' \\
&= 1 + (s-1)\boldsymbol{\xi}(\boldsymbol{I} - s\boldsymbol{N})^{-1}\boldsymbol{1}'.
\end{aligned}
$$

This proves the first part of the theorem. The second part, the mean of the waiting time $W(\Lambda)$, is an immediate consequence of

$$\varphi_W^{(1)}(s)|_{s=1} = \frac{d}{ds}\left\{1 + (s-1)\boldsymbol{\xi}(\boldsymbol{I} - s\boldsymbol{N})^{-1}\boldsymbol{1}'\right\}|_{s=1} = \boldsymbol{\xi}(\boldsymbol{I} - \boldsymbol{N})^{-1}\boldsymbol{1}'. \qquad \square$$

In the following, we extend Theorem 5.3 to the case where Λ is a compound pattern.

Theorem 5.4   *If $\Lambda = \cup_{i=1}^{l} \Lambda_i$ is a compound pattern of l distinct simple patterns, then Eqs. (5.24) and (5.25) hold.*

Proof.   For a compound pattern $\Lambda = \cup_{i=1}^{l} \Lambda_i$, we have

$$P(W(\Lambda) = n) = \sum_{i=1}^{l} P(W(\Lambda) = n \text{ and } W(\Lambda_i) = n), \qquad (5.26)$$

for $n = 1, 2, \cdots, \infty$. For each $i = 1, \cdots, l$, the generating function $\phi_i(s)$ of the sequence $\{P(W(\Lambda) = n \text{ and } W(\Lambda_i) = n)\}_{i=1}^{\infty}$ is given by

$$
\begin{aligned}
\phi_i(s) &= \sum_{n=1}^{\infty} s^n P(W(\Lambda) = n \text{ and } W(\Lambda_i) = n) \\
&= \sum_{n=1}^{\infty} s^n P(Y_n = \alpha_i, Y_{n-1} \notin A, \cdots, Y_1 \notin A) \\
&= \sum_{n=1}^{\infty} s^n \boldsymbol{\xi} \boldsymbol{N}^{n-1} \boldsymbol{C}_i \\
&= s\boldsymbol{\xi}(\boldsymbol{I} - s\boldsymbol{N})^{-1} \boldsymbol{C}_i. \qquad (5.27)
\end{aligned}
$$

In view of Eq. (5.26), we have

$$\varphi_w(s) = \sum_{i=1}^{l} \phi_i(s) = \sum_{i=1}^{l} s\boldsymbol{\xi}(\boldsymbol{I} - s\boldsymbol{N})^{-1}\boldsymbol{C}_i. \qquad (5.28)$$

Since

$$(\boldsymbol{I} - \boldsymbol{N})\boldsymbol{1}' = \sum_{i=1}^{l} \boldsymbol{C}_i,$$

and

$$s\boldsymbol{\xi}(\boldsymbol{I} - s\boldsymbol{N})^{-1}\boldsymbol{N}\boldsymbol{1}' = \boldsymbol{\xi}(\boldsymbol{I} - s\boldsymbol{N})^{-1}\boldsymbol{1}' - 1,$$

Eqs. (5.24) and (5.25) follow immediately. $\square$

Example 5.3   Let $\{X_t\}$ be a sequence of Bernoulli trials and $\Lambda = \Lambda_1 \cup \Lambda_2$ be a compound pattern composed of $\Lambda_1 = SS$ and $\Lambda_2 = FF$. The imbedded

Markov chain $\{Y_t\}$ for the waiting time $W(\Lambda)$ has the state space $\Omega = \{S, F, \alpha_1, \alpha_2\}$ and the transition probability matrix

$$
M = \begin{array}{c} S \\ F \\ \alpha_1 \\ \alpha_2 \end{array}
\left(\begin{array}{cc|cc}
0 & q & p & 0 \\
p & 0 & 0 & q \\ \hline
0 & 0 & 1 & 0 \\
0 & 0 & 0 & 1
\end{array} \right)
= \left(\begin{array}{c|c} N & C \\ \hline 0 & I \end{array} \right),
$$

with an initial distribution for Y_1 of $(p, q, 0, 0)$. Note that for convenience in obtaining the inverse $(I - sN)^{-1}$ below by hand, here we use the initial distribution of Y_1 instead of Y_0 (at the dummy state $\emptyset$).

In the following, we obtain the probability generating function $W(\Lambda)$ by two methods: the recursive method introduced in Section 3.7, and Theorem 5.4.

A simple manipulation via Theorem 5.2(iii) yields the following recursive equation:

$$
\begin{aligned}
\phi_1(s) &= \sum_{n=2}^{\infty} s^n \boldsymbol{\xi}_1 N^{n-2} C_1 \\
&= p^2 s^2 + p^2 q s^3 + pq s^2 \phi_1(s),
\end{aligned}
$$

where $\boldsymbol{\xi}_1 = (p, q)$. Hence

$$
\phi_1(s) = \frac{p^2 s^2 + p^2 q s^3}{1 - pq s^2}.
$$

Similarly, we have

$$
\phi_2(s) = \frac{q^2 s^2 + q^2 p s^3}{1 - pq s^2}.
$$

It follows from Eq. (5.28) that

$$
\varphi_W(s) = \frac{(p^2 + q^2) s^2 + pq s^3}{1 - pq s^2}.
$$

The above result can also be obtained from the definition of the generating function and Theorem 5.2(ii), as well as directly from Theorem 5.4:

$$
\begin{aligned}
\varphi_W(s) &= \sum_{n=2}^{\infty} s^n \boldsymbol{\xi}_1 N^{n-2} (I - N) \mathbf{1}' \\
&= s^2 \boldsymbol{\xi}_1 (I - sN)^{-1} (I - N) \mathbf{1}'
\end{aligned}
$$

$$= s^2(p,q) \begin{pmatrix} 1 & -sq \\ -sp & 1 \end{pmatrix}^{-1} \begin{pmatrix} p \\ q \end{pmatrix}$$

$$= \frac{(p^2 + q^2)s^2 + pqs^3}{1 - pqs^2}.$$

$$\diamond$$

5.5 Mean of Waiting Time $W(\Lambda)$

In Theorems 5.3 and 5.4, it was shown that the mean waiting time of a compound pattern is $EW(\Lambda) = \boldsymbol{\xi}(\boldsymbol{I} - \boldsymbol{N})^{-1}\boldsymbol{1}'$. This formula is efficient only if the size of $\boldsymbol{N}$ is not too large. To find an analytic form of $EW(\Lambda)$ using the above formula is rather hard. In view of this complexity, we provide the following alternative result, which often yields an analytic form for the mean waiting time $EW(\Lambda)$ with less effort.

For the imbedded Markov chain $\{Y_t\}$ of the waiting-time random variable $W(\Lambda)$, the state space and the transition probability matrix can always be relabeled as $\Omega = \{1, 2, \cdots, w, \alpha_1, \cdots, \alpha_l\}$ and

$$\boldsymbol{M} = \left(\begin{array}{c|c} \boldsymbol{N} & \boldsymbol{C} \\ \hline \boldsymbol{0} & \boldsymbol{I} \end{array} \right),$$

where w denotes the number of non-absorbing states.

Theorem 5.5

$$EW(\Lambda) = S_1 + \cdots + S_w, \tag{5.29}$$

where $\{S_1, \cdots, S_w\}$ is the solution of the following simultaneous equations:

$$S_i = \boldsymbol{\xi}\boldsymbol{e}_i' + (S_1, \cdots, S_w)\boldsymbol{N}\boldsymbol{e}_i', \quad i = 1, \cdots, w, \tag{5.30}$$

and $\boldsymbol{N}\boldsymbol{e}_i'$ is the i-th column vector of the matrix $\boldsymbol{N}$.

Proof. It follows from the definition of $EW(\Lambda)$ and Theorem 5.2(ii) that

$$\begin{aligned} EW(\Lambda) &= \sum_{n=1}^{\infty} P(W(\Lambda) \geq n) \\ &= \sum_{n=1}^{\infty} \boldsymbol{\xi}\boldsymbol{N}^{n-1}\boldsymbol{1}' \end{aligned}$$

$$= \sum_{n=1}^{\infty}\sum_{i=1}^{w} \boldsymbol{\xi N}^{n-1}\boldsymbol{e}_i'$$

$$= \sum_{i=1}^{w}\sum_{n=1}^{\infty} \boldsymbol{\xi N}^{n-1}\boldsymbol{e}_i'$$

$$= S_1 + \cdots + S_w,$$

where $S_i = \sum_{n=1}^{\infty} \boldsymbol{\xi N}^{n-1}\boldsymbol{e}_i'$, $i = 1, 2, \cdots, w$. Note that

$$\boldsymbol{\xi N}^{n-1}\boldsymbol{e}_i' = \boldsymbol{\xi N}^{n-2}\boldsymbol{N}\boldsymbol{e}_i'$$

$$= \sum_{j=1}^{w} p_{ji}\boldsymbol{\xi N}^{n-2}\boldsymbol{e}_j',$$

and

$$S_i = \sum_{n=1}^{\infty} \boldsymbol{\xi N}^{n-1}\boldsymbol{e}_i'$$

$$= \boldsymbol{\xi e}_i' + \sum_{n=2}^{\infty} \boldsymbol{\xi N}^{n-2}\sum_{j=1}^{w} p_{ji}\boldsymbol{e}_j'$$

$$= \boldsymbol{\xi e}_i' + \sum_{j=1}^{w} p_{ji}\sum_{n=2}^{\infty} \boldsymbol{\xi N}^{n-2}\boldsymbol{e}_j'$$

$$= \boldsymbol{\xi e}_i' + \sum_{j=1}^{w} p_{ji}S_j$$

$$= \boldsymbol{\xi e}_i' + (S_1, \cdots, S_w)\boldsymbol{N}\boldsymbol{e}_i'.$$

This completes the proof. □

Example 5.4 Let $\{X_t\}$ be a sequence of Bernoulli trials with $P(X = S) = p$ and $P(X = F) = q$, and let $W(\Lambda)$ be the waiting time for the pattern $\Lambda = S\cdots S$ of k consecutive successes. The imbedded Markov chain $\{Y_t\}$, defined on the state space

$$\Omega = \{F, S, SS, \cdots, \underbrace{S\cdots S}_{k-1}, \alpha\},$$

has the transition probability matrix of Eq. (3.28):

$$M = \begin{pmatrix} q & p & 0 & \cdots & \cdots & 0 & 0 \\ q & 0 & p & 0 & \cdots & 0 & 0 \\ \vdots & & \ddots & & & & \vdots \\ \vdots & & & \ddots & & & \vdots \\ \vdots & & & & \ddots & & \vdots \\ q & & & & & 0 & p \\ 0 & \cdot & & & & 0 & 1 \end{pmatrix} = \left(\begin{array}{c|c} N & C \\ \hline 0 & I \end{array} \right).$$

For *i.i.d.* cases, we set the initial distribution to $Y_0 = F$ with probability one, *i.e.* $\boldsymbol{\xi} = (1, 0, \cdots, 0)_{1 \times k}$. Hence it follows that (here $w = k$)

$$S_1 = \sum_{n=1}^{\infty} \boldsymbol{\xi} N^{n-1} e_1' = 1 + q(S_1 + S_2 + \cdots + S_k)$$

and

$$S_i = \sum_{n=1}^{\infty} \boldsymbol{\xi} N^{n-1} e_i' = p S_{i-1}, \quad i = 2, \cdots, k.$$

This yields the mean waiting time

$$EW(\Lambda) = \frac{1 + p + \cdots + p^{k-1}}{1 - q - pq - \cdots - p^{k-1}q} = \frac{1}{p^k} + \frac{1}{p^{k-1}} + \cdots + \frac{1}{p}.$$

$$\diamond$$

Example 5.5   For the pattern $\Lambda = SFS$ in Bernoulli trials, the imbedded Markov chain $\{Y_t\}$ of the waiting time $W(\Lambda)$ has the state space $\Omega = \{F, S, SF, \alpha\}$ and the transition matrix

$$M = \begin{array}{c} F \\ S \\ SF \\ \alpha \end{array} \left(\begin{array}{ccc|c} q & p & 0 & 0 \\ 0 & p & q & 0 \\ q & 0 & 0 & p \\ \hline 0 & 0 & 0 & 1 \end{array} \right).$$

Given $\boldsymbol{\xi} = (1, 0, 0)$, it follows from Eq. (5.30) that

$$\begin{aligned} S_1 &= 1 + q(S_1 + S_3), \\ S_2 &= p(S_1 + S_2), \\ S_3 &= q S_2. \end{aligned}$$

Solving the above three simultaneous equations yields

$$EW(\Lambda) = \frac{1}{p} + \frac{1}{pq} + \frac{1}{p^2}.$$

$\diamond$

Example 5.6 Let $\Lambda = \Lambda_1 \cup \Lambda_2$ be a compound pattern with $\Lambda_1 = FFF$ and $\Lambda_2 = SSS$. The imbedded Markov chain associated with the waiting time $W(\Lambda)$ in Bernoulli trials has the state space $\Omega = \{F, S, FF, SS, \alpha_1, \alpha_2\}$ and the transition probability matrix

$$M = \begin{array}{c} F \\ S \\ FF \\ SS \\ \alpha_1 \\ \alpha_2 \end{array} \left(\begin{array}{cccc:cc} 0 & p & q & 0 & 0 & 0 \\ q & 0 & 0 & p & 0 & 0 \\ 0 & p & 0 & 0 & q & 0 \\ q & 0 & 0 & 0 & 0 & p \\ \hdashline 0 & 0 & 0 & 0 & 1 & 0 \\ 0 & 0 & 0 & 0 & 0 & 1 \end{array} \right).$$

Choosing the initial probability $\boldsymbol{\xi}_1 = (q, p, 0, 0)$ for Y_1, we then have the following system of simultaneous equations:

$$\begin{aligned} S_1 &= q + q(S_2 + S_4), \\ S_2 &= p + p(S_1 + S_3), \\ S_3 &= qS_1, \\ S_4 &= pS_2. \end{aligned}$$

Basic manipulation yields

$$\begin{aligned} E[W(\Lambda)|\boldsymbol{\xi}_1 = (q,p,0,0)] &= S_1 + S_2 + S_3 + S_4 \\ &= \frac{2(1 + pq + p^2 q^2)}{1 - 2pq - p^2 q^2}. \end{aligned}$$

$\diamond$

Although all of the examples above are for *i.i.d.* cases, the results of Theorem 5.5 also hold for multi-state homogeneous Markov-dependent trials with some simple modification to the transition probability matrix $\boldsymbol{M}$.

The connection between Theorem 5.4 and Theorem 5.5 for finding $EW(\Lambda)$ is straightforward. Since $(S_1, \cdots, S_w)$ is the solution of the simultaneous equations $S_i = \boldsymbol{\xi}\boldsymbol{e}_i' + (S_1, \cdots, S_w)\boldsymbol{N}\boldsymbol{e}_i'$, $i = 1, \cdots, w$, it follows that $(S_1, S_2,$

$\cdots, S_w) = \boldsymbol{\xi} \, (\boldsymbol{I} - \boldsymbol{N})^{-1}$. Hence we have

$$EW(\Lambda) = S_1 + \cdots + S_w = \boldsymbol{\xi}(\boldsymbol{I} - \boldsymbol{N})^{-1}\mathbf{1}'.$$

Further, since the expected waiting time depends on $\boldsymbol{\xi}_0 = (\boldsymbol{\xi} : \mathbf{0})$, the initial distribution of the imbedded Markov chain $\{Y_t\}$, the latter should be set according to the particular application, especially when $\{X_t\}$ is a Markov chain and Λ is a compound pattern. If the dummy state $\{\emptyset\}$ is added in Example 5.6 with $P(Y_0 = \emptyset) \equiv 1$, for instance, then the imbedded Markov chain $\{Y_t\}$ is defined on the state space $\Omega^{\emptyset} = \{\emptyset, F, S, FF, SS, \alpha_1, \alpha_2\}$ with transition probability matrix

$$\boldsymbol{M}^{\emptyset} = \begin{matrix} \emptyset \\ F \\ S \\ FF \\ SS \\ \alpha_1 \\ \alpha_2 \end{matrix} \left(\begin{array}{ccccc|cc} 0 & q & p & 0 & 0 & 0 & 0 \\ 0 & 0 & p & q & 0 & 0 & 0 \\ 0 & q & 0 & 0 & p & 0 & 0 \\ 0 & 0 & p & 0 & 0 & q & 0 \\ 0 & q & 0 & 0 & 0 & 0 & p \\ \hline 0 & 0 & 0 & 0 & 0 & 1 & 0 \\ 0 & 0 & 0 & 0 & 0 & 0 & 1 \end{array} \right) = \left(\begin{array}{c|c} \boldsymbol{N}^{\emptyset} & \boldsymbol{C}^{\emptyset} \\ \hline \boldsymbol{O} & \boldsymbol{I} \end{array} \right),$$

and the mean waiting time of $W(\Lambda)$ equals

$$E[W(\Lambda)|\boldsymbol{\xi} = (1,0,0,0,0)] = \frac{3 + p^2 q^2}{1 - 2pq - p^2 q^2}.$$

Note that

$$E[W(\Lambda)|\boldsymbol{\xi} = (1,0,0,0,0)] - E[(W(\Lambda)|\boldsymbol{\xi}_1 = (q,p,0,0)] \equiv 1.$$

This is due to the fact that an additional step is required by starting at Y_0 instead of Y_1, and we have the relationship

$$\boldsymbol{\xi}\boldsymbol{N}^{\emptyset} = (1,0,0,0,0)\boldsymbol{N}^{\emptyset} = (0,q,p,0,0) = (0 : \boldsymbol{\xi}_1).$$

5.6   More About Generating Functions

Obtaining the explicit form of the probability generating function $\varphi_W(s)$ through Theorems 5.3 and 5.4 is often a hard task because it requires finding the analytic form of the inverse of the matrix $(\boldsymbol{I} - s\boldsymbol{N})$. Parallel to Theorem 5.5, the analytic form of the probability generating function $\varphi_W(s)$ can be written in terms of $\Phi_W(s) = \sum_{n=1}^{\infty} s^n P(W(\Lambda) \geq n)$, the probability generating function of cumulative probabilities $\{P(W(\Lambda) \geq n)\}$.

Theorem 5.6 *Assume that the imbedded Markov chain $\{Y_t\}$ correspond-*
ing to the waiting time $W(\Lambda)$ is defined on the state space $\Omega = \{1, 2, \cdots, w, \alpha_1,$
$\cdots, \alpha_l\}$ with a transition probability matrix $\mathbf{M}$ having the form of Eq. (5.13).
Then

(i)

$$\Phi_W(s) = \phi_1(s) + \cdots + \phi_w(s), \tag{5.31}$$

where $(\phi_1(s), \phi_2(s), \cdots, \phi_w(s))$ is the solution of the simultaneous
equations

$$\phi_i(s) = s\boldsymbol{\xi}e_i' + s(\phi_1(s), \cdots, \phi_w(s))\mathbf{N}e_i', \tag{5.32}$$

for $i = 1, \cdots, w$, and

(ii)

$$\varphi_W(s) = 1 + \Phi_W(s) - \frac{1}{s}\Phi_W(s). \tag{5.33}$$

Proof. The proof of part (i) is the same as the proof of Theorem 5.5. We
leave this to the reader. The second part of the theorem follows from the
definitions and

$$
\begin{aligned}
\varphi_W(s) &= \sum_{n=1}^{\infty} s^n \boldsymbol{\xi} \mathbf{N}^{n-1}(\mathbf{I} - \mathbf{N})\mathbf{1}' \\
&= \sum_{n=1}^{\infty} s^n \boldsymbol{\xi} \mathbf{N}^{n-1}\mathbf{1}' - \sum_{n=1}^{\infty} s^n \boldsymbol{\xi} \mathbf{N}^n \mathbf{1}' \\
&= \sum_{n=1}^{\infty} s^n \boldsymbol{\xi} \mathbf{N}^{n-1}\mathbf{1}' - \frac{1}{s}\sum_{n=0}^{\infty} s^{n+1} \mathbf{N}^n \mathbf{1}' + 1 \\
&= 1 + \Phi_W(s) - \frac{1}{s}\Phi_W(s).
\end{aligned}
$$

This completes the proof. □

Example 5.7 Consider $\Lambda = SS$ in Bernoulli trials. The imbedded
Markov chain $\{Y_t\}$ for $W(\Lambda)$ has the state space $\Omega = \{F, S, \alpha\}$ and the
transition probability matrix

$$
\mathbf{M} = \begin{array}{c} F \\ S \\ \alpha \end{array} \left(\begin{array}{cc|c} q & p & 0 \\ q & 0 & p \\ \hline 0 & 0 & 1 \end{array} \right).
$$

Given $\boldsymbol{\xi} = (1,0)$, it follows that

$$\phi_1(s) = s + qs(\phi_1(s) + \phi_2(s)),$$

and

$$\phi_2(s) = ps\phi_1(s).$$

Simple manipulation yields

$$\Phi_W(s) = \frac{s + ps^2}{1 - qs - qps^2}$$

and

$$
\begin{aligned}
\varphi_W(s) &= 1 + \Phi_W(s) - \frac{1}{s}\Phi_W(s) \\
&= \frac{p^2 s^2}{1 - qs - qps^2} = \frac{p^2 s^2 (1 - ps)}{1 - s + qp^2 s^3}.
\end{aligned}
$$

$\diamond$

The probability generating function $\varphi_W(s)$ can also be obtained by using forward multiplication in the following way. Note that the initial distribution can always be written as $\boldsymbol{\xi} = \sum_{i=1}^{w} \xi_i \boldsymbol{e}_i$. It follows from the definition of the probability generating function that

$$\varphi_W(s) = \sum_{n=1}^{\infty} s^n \boldsymbol{\xi} \boldsymbol{N}^{n-1}(\boldsymbol{I} - \boldsymbol{N})\boldsymbol{1}' = \sum_{n=1}^{w} \xi_i \varphi_i(s), \qquad (5.34)$$

where, for $i = 1, \cdots, w$,

$$\varphi_i(s) = \sum_{i=1}^{\infty} s^n \boldsymbol{e}_i \boldsymbol{N}^{n-1}(\boldsymbol{I} - \boldsymbol{N})\boldsymbol{1}' \qquad (5.35)$$

is the probability generating function with initial distribution $\boldsymbol{e}_i$. For each i, it follows from this first-step analysis that

$$
\begin{aligned}
\varphi_i(s) &= \sum_{n=1}^{\infty} s^n \boldsymbol{e}_i \boldsymbol{N}^{n-1}(\boldsymbol{I} - \boldsymbol{N})\boldsymbol{1}' \\
&= s\boldsymbol{e}_i(\boldsymbol{I} - \boldsymbol{N})\boldsymbol{1}' + \sum_{n=2}^{\infty} s^n \boldsymbol{e}_i \boldsymbol{N}^{n-1}(\boldsymbol{I} - \boldsymbol{N})\boldsymbol{1}' \\
&= s\boldsymbol{e}_i(\boldsymbol{I} - \boldsymbol{N})\boldsymbol{1}' + s\sum_{n=2}^{\infty} s^{n-1}(\boldsymbol{e}_i \boldsymbol{N})\boldsymbol{N}^{n-2}(\boldsymbol{I} - \boldsymbol{N})\boldsymbol{1}'
\end{aligned}
$$

$$
\begin{aligned}
&= se_i(\boldsymbol{I}-\boldsymbol{N})\boldsymbol{1}' + s\sum_{n=2}^{\infty}\sum_{j=1}^{w} s^{n-1}p_{ij}e_j\boldsymbol{N}^{n-2}(\boldsymbol{I}-\boldsymbol{N})\boldsymbol{1}' \\
&= se_i(\boldsymbol{I}-\boldsymbol{N})\boldsymbol{1}' + s\sum_{j=1}^{w} p_{ij}\varphi_j(s). \tag{5.36}
\end{aligned}
$$

Equations (5.34), (5.35), and (5.36) yield the following theorem.

Theorem 5.7 *Assume that the imbedded Markov chain $\{Y_t\}$ correspond-ing to the waiting time $W(\Lambda)$ is defined on the state space $\Omega = \{1, 2, \cdots, w,$ $\alpha_1, \cdots, \alpha_l\}$ with a transition probability matrix $\boldsymbol{M}$ having the form of Eq. (5.13). Then given $\boldsymbol{\xi}_0 = (\boldsymbol{\xi} : \boldsymbol{0})$, the probability generating function of $W(\Lambda)$ is*

$$\varphi_W(s) = \boldsymbol{\xi}(\varphi_1(s), \cdots, \varphi_w(s))',$$

where $(\varphi_1(s), \cdots, \varphi_w(s))$ is the solution of the simultaneous equations

$$\varphi_i(s) = se_i(\boldsymbol{I}-\boldsymbol{N})\boldsymbol{1}' + se_i\boldsymbol{N}(\varphi_1(s), \cdots, \varphi_w(s))',$$

for $i = 1, \cdots, w$, and $e_i\boldsymbol{N}$ is the i-th row of the matrix $\boldsymbol{N}$.

For the special case $\boldsymbol{\xi} = (1, 0, \cdots, 0)$, Aki and Hirano (1999, 2000), Aki (1999), and Han and Aki (1998, 2000a) refer to the probability generating function $\varphi_W(s) = \varphi_1(s)$ as the unconditional probability generating func-tion, and to $\varphi_i(s)$, $i = 2, 3, \cdots, w$, as the conditional probability generating functions. In this case, the distribution of the waiting-time random variable $W(\Lambda)$ can be obtained through

$$P(W(\Lambda) = n) = \frac{1}{n!}(\frac{d}{ds})^n\varphi_1(s)|_{s=0}.$$

They also refer to this approach for obtaining the distribution of the waiting-time random variable $W(\Lambda)$ as the conditional probability generating func-tion technique.

In addition to the above *i.i.d.* examples, we would like to give an ad-ditional example to illustrate that our results are also applicable for the case where $\{X_t\}$ is a two-state homogeneous Markov chain with transition probability matrix

$$
\boldsymbol{A} = \begin{pmatrix} p_{FF} & p_{FS} \\ p_{SF} & p_{SS} \end{pmatrix}.
$$

Example 5.8   Let us consider the waiting-time random variable $W(\Lambda)$ of a compound pattern $\Lambda = \Lambda_1 \cup \Lambda_2$, where $\Lambda_1 = SS$ and $\Lambda_2 = FF$. Given the initial probabilities $P(X_1 = S) = p$ and $P(X_1 = F) = q = 1 - p$, the imbedded Markov chain $\{Y_t\}$ associated with the waiting time $W(\Lambda)$ has the state space $\Omega = \{\emptyset, S, F, \alpha_1, \alpha_2\}$ and the transition probability matrix

$$M = \begin{matrix} \emptyset \\ S \\ F \\ \alpha_1 \\ \alpha_2 \end{matrix} \begin{pmatrix} 0 & p & q & 0 & 0 \\ 0 & 0 & p_{SF} & p_{SS} & 0 \\ 0 & p_{FS} & 0 & 0 & p_{FF} \\ 0 & 0 & 0 & 1 & 0 \\ 0 & 0 & 0 & 0 & 1 \end{pmatrix} = \left(\begin{array}{c|c} N & C \\ \hline 0 & I \end{array} \right).$$

Given the initial distribution $\boldsymbol{\xi} = (1, 0, 0)$, Theorems 5.5 and 5.6 yield the mean waiting time and the probability generating function, respectively, as

$$E[W(\Lambda)] = 1 + \frac{p + qp_{FS}}{1 - p_{SF}p_{FS}} + \frac{q + pp_{SF}}{1 - p_{SF}p_{FS}}$$

and

$$\varphi_W(s) = p_{SS} \frac{ps^2 + qp_{FS}s^3}{1 - p_{SF}p_{FS}s^2} + p_{FF} \frac{qs^2 + pp_{SF}s^3}{1 - p_{SF}p_{FS}s^2}.$$

$$\diamondsuit$$

5.7   Spectrum Analysis and Large Deviation Approximation

Let $\lambda_{[1]}, \lambda_{[2]}, \cdots, \lambda_{[w]}$, be the ordered eigenvalues (in the sense of $|\lambda_{[1]}| \geq |\lambda_{[2]}| \geq \cdots \geq |\lambda_{[w]}|$) of the essential transition probability submatrix N, and let $\boldsymbol{\eta}_1', \boldsymbol{\eta}_2', \cdots, \boldsymbol{\eta}_w'$, be the column eigenvectors corresponding to the eigenvalues $\lambda_{[i]}$, in the sense that $N\boldsymbol{\eta}_i' = \lambda_{[i]}\boldsymbol{\eta}_i'$, for all $i = 1, \cdots, w$. Since N is a substochastic matrix, it is known by the Perron-Frobenius Theorem (Seneta 1981) that the largest eigenvalue $\lambda_{[1]}$ is unique and a real number between zero and one $(0 < \lambda_{[1]} < 1)$. Since $\mathbf{1}'$ can always be written as a linear combination of the eigenvectors $\boldsymbol{\eta}_i'$, i.e.,

$$\mathbf{1}' = \sum_{i=1}^{w} a_i \boldsymbol{\eta}_i', \qquad (5.37)$$

we obtain the following theorem.

Theorem 5.8 *If the transition probability matrix $\boldsymbol{M}$ corresponding to the waiting time $W(\Lambda)$ has the form of Eq. (5.13), then*

(i)

$$P(W(\Lambda) \geq n) = \sum_{i=1}^{w} C_i \lambda_{[i]}^{n-1}, \qquad (5.38)$$

where $C_i = a_i \boldsymbol{\xi} \boldsymbol{\eta}_i'$, $i = 1, \cdots, w$,

(ii)

$$\Phi_W(s) = \sum_{n=1}^{\infty} s^n P(W(\Lambda) \geq n) = \sum_{i=1}^{w} \frac{C_i s}{1 - s\lambda_{[i]}}$$

and $\Phi_W(s)$ exists for $|s| < 1/\lambda_{[1]}$, and

(iii)

$$\varphi_W(s) = 1 + (1 - \frac{1}{s}) \sum_{i=1}^{w} \frac{C_i s}{1 - s\lambda_{[i]}}, \quad |s| < 1/\lambda_{[1]}. \qquad (5.39)$$

Proof. It follows from Eq. (5.37) that, for every n,

$$
\begin{aligned}
P(W(\Lambda) \geq n) &= \boldsymbol{\xi} \boldsymbol{N}^{n-1} \boldsymbol{1}' = \boldsymbol{\xi} \boldsymbol{N}^{n-1} \sum_{i=1}^{w} a_i \boldsymbol{\eta}_i' \\
&= \sum_{i=1}^{w} a_i \boldsymbol{\xi} \boldsymbol{N}^{n-1} \boldsymbol{\eta}_i' \\
&= \sum_{i=1}^{w} C_i \lambda_{[i]}^{n-1}. \qquad (5.40)
\end{aligned}
$$

This completes the proof of part (i). The result (ii) follows immediately from the definition of $\Phi_W(s)$ and Eq. (5.40). Note that $\Phi_W(s)$ exists if and only if $\sum_{n=1}^{\infty} \lambda_{[i]}^n s^n < \infty$, for all $i = 1, \cdots, w$. Since all the series are power series and $\lambda_{[1]}$ is the largest eigenvalue ($0 < \lambda_{[1]} < 1$), $\Phi_W(s)$ exists if and only if $|s| < 1/\lambda_{[1]}$, and summing the series gives

$$\Phi_W(s) = \sum_{i=1}^{w} \frac{C_i s}{1 - s\lambda_{[i]}}. \qquad (5.41)$$

The result (iii) follows from Theorem 5.6(ii) and Eq. (5.41). $\qquad \square$

The largest eigenvalue $\lambda_{[1]}$ is always less than 1 $(0 < \lambda_{[1]} < 1)$, even though its value varies from problem to problem. Hence all the probability generating functions and the cumulative generating functions always exist in the range $|s| \leq 1$.

Further, it follows from Eqs. (5.39) and (5.41) that

$$EW(\Lambda) = \frac{d}{ds}\varphi_W(s)|_{s=1} = \Phi_W(1) = \sum_{i=1}^{w} \frac{C_i}{1 - \lambda_{[i]}}. \tag{5.42}$$

Theorem 5.9 *As $n \to \infty$, the tail probability $P(W(\Lambda) \geq n)$ converges to zero exponentially in the following sense:*

$$\lim_{n\to\infty} \frac{1}{n} \log P(W(\Lambda) \geq n) = -\beta(\Lambda), \tag{5.43}$$

where $\beta(\Lambda) = -\log \lambda_{[1]}$.

The tail probability of the general geometric distribution is determined by the largest eigenvalue of N, and, for large n, Eq. (5.43) provides a large deviation approximation for the tail probability; *i.e.*

$$P(W(\Lambda) \geq n) = \exp\{-n\beta(\Lambda)[1 + o(1)]\} \sim \exp\{n\log\lambda_{[1]}\}. \tag{5.44}$$

Proof. It can be shown that $\sum_{i=1}^{w} a_i \boldsymbol{\xi}\boldsymbol{\eta}_i' = 1$ and $C_1 = a_1\boldsymbol{\xi}\boldsymbol{\eta}_1' > 0$. Since $\lambda_{[1]}$ is unique and $\lambda_{[1]} > |\lambda_{[i]}|$, for all $i = 2, \cdots, w$, by Eq. (5.38) we have the following inequality

$$C_1\lambda_{[1]}^{n-1}\left\{1 - \sum_{i=2}^{w}\left|\frac{C_i}{C_1}\right|\left|\frac{\lambda_{[i]}}{\lambda_{[1]}}\right|^{n-1}\right\} \leq P(W(\Lambda) \geq n)$$

$$\leq C_1\lambda_{[1]}^{n-1}\left\{1 + \sum_{i=2}^{w}\left|\frac{C_i}{C_1}\right|\left|\frac{\lambda_{[i]}}{\lambda_{[1]}}\right|^{n-1}\right\}.$$

Given an arbitrarily small number $\epsilon > 0$, since $|\lambda_{[i]}/\lambda_{[1]}|^{n-1} \to 0$ as $n \to \infty$ for all $i = 2, \cdots, w$, there exists a number n_0 such that for all $n \geq n_0$,

$$C_1\lambda_{[1]}^{n-1}\{1 - \epsilon\} \leq P(W(\Lambda) \geq n) \leq C_1\lambda_{[1]}^{n-1}\{1 + \epsilon\}. \tag{5.45}$$

Taking the log of both sides of Eq. (5.45) and dividing by n, then as $n \to \infty$, we have

$$\lim_{n\to\infty} \frac{1}{n} \log P(W(\Lambda) \geq n) = \log\lambda_{[1]}.$$

This completes the proof. $\qquad\qquad\qquad\qquad\qquad\qquad\qquad\qquad\square$

In the language of large deviation theory, we can say that the tail probability $P(W(\Lambda) \geq n)$ obeys the first-order large deviation principle in the sense of Eq. (5.43). The positive constant $\beta(\Lambda) = -\log \lambda_{[1]}$ is often referred to as the exponential rate of $P(W(\Lambda) \geq n)$ tending to zero. Numerical examples for the large deviation approximation in Eq. (5.44) are provided in Section 7.2.

Following Theorem 5.9, we can prove a slightly stronger result.

Theorem 5.10

$$\lim_{n \to \infty} \frac{P(W(\Lambda) \geq n)}{C_1 \lambda_{[1]}^{n-1}} = 1. \tag{5.46}$$

Proof. It follows from Eq. (5.38) that there exists a constant c such that

$$1 - c \left| \frac{\lambda_{[2]}}{\lambda_{[1]}} \right|^{n-1} \leq \frac{P(W(\Lambda) \geq n)}{C_1 \lambda_{[1]}^{n-1}} \leq 1 + c \left| \frac{\lambda_{[2]}}{\lambda_{[1]}} \right|^{n-1}. \tag{5.47}$$

Since $\lambda_{[1]} > |\lambda_{[2]}|$, the result is an immediate consequence of the above inequality by letting n tend to infinity. This yields, for large n, that the tail probability of $W(\Lambda)$ can be approximated by

$$P(W(\Lambda) \geq n) \sim C_1 \lambda_{[1]}^{n-1}.$$

$\square$

5.8 Probability Generating Function of $W(r, \Lambda)$

The waiting-time random variable $W(r, \Lambda)$ can be viewed as the sum of r waiting times of the pattern Λ under non-overlap counting:

$$W(r, \Lambda) = W_1(\Lambda) + \cdots + W_r(\Lambda).$$

It is easy to see that $\{W_i(\Lambda)\}$ are *i.i.d.* (*i.e.* $W(\Lambda) \stackrel{d}{=} W_i(\Lambda)$ for all i) if $\{X_t\}$ are *i.i.d.* multi-state trials, and the counting is non-overlapping. Then the probability generating function of $W(r, \Lambda)$ is $\varphi_{W(r,\Lambda)}(s) = (\varphi_W(s))^r$.

Under overlap counting for *i.i.d.* trials, the waiting time for r patterns Λ, $W^o(r, \Lambda)$ (superscript "o" indicates overlap counting), can also be written as a sum of the waiting times of the first, second, and so on, occurrences of the simple pattern Λ:

$$W^o(r, \Lambda) = W_1(\Lambda) + W_2^o(\Lambda) + \cdots + W_r^o(\Lambda). \tag{5.48}$$

Let E^o be the ending block with respect to overlap counting (see definition of E^o in Section 4.3). Given $\boldsymbol{\xi}(E^o)$, the random variables $W_i^o(\Lambda)$, $i = 2, \cdots, r$, are conditionally independent and identically distributed, and we denote these variables by $W^o(\Lambda)$. Note that the imbedded Markov chain $\{Y_t\}$ for $W^o(\Lambda)$ has the same transition probability matrix as the imbedded Markov chain for the waiting time $W(\Lambda)$ (under non-overlap counting), and has initial distribution $\boldsymbol{\xi}(E^o)$, where $\boldsymbol{\xi}(E^o)$ is a unit vector with one at state E^o and zero elsewhere. Hence the probability generating function for $W^o(\Lambda)$ is

$$\varphi_{W^o}(s) = \sum_{n=1}^{\infty} s^n \boldsymbol{\xi}(E^o) \boldsymbol{N}^{n-1} (\boldsymbol{I} - \boldsymbol{N}) \mathbf{1}'. \tag{5.49}$$

It follows that the probability generating function of $W^o(r, \Lambda)$ is

$$\varphi_{W^o(r,\Lambda)}(s) = \varphi_W(s)(\varphi_{W^o}(s))^{r-1}. \tag{5.50}$$

If $\{X_t\}$ is a sequence of homogeneous Markov-dependent trials, Λ is a simple pattern, and counting is non-overlapping, then the probability generating function of $W(r, \Lambda)$ has the same form as Eq. (5.50), except that the initial distribution for the second and subsequent patterns, $\boldsymbol{\xi}(\emptyset_e)$, is degenerated at the state $\emptyset_e$, the initial state for counting the second pattern Λ when e is the last element of pattern Λ. The transition probability from $\emptyset_e$ to $v \in \Omega$ is given by

$$p_{\emptyset_e, v} = \begin{cases} p_{eb} & \text{if } v = < \emptyset_e, b >_\Omega \text{ with } b \in \mathbb{S} \\ 0 & \text{otherwise.} \end{cases}$$

Further, Eq. (5.50) also works for the case where $\{X_t\}$ is a sequence of homogeneous Markov-dependent trials, Λ is a simple pattern, and counting is overlapping.

To illustrate Theorem 5.6 and Eq. (5.50), we provide the following example.

Example 5.9   Assume $\{X_t\}$ is a sequence of Bernoulli trials and $\Lambda = SS$. It follows that Λ yields $E^o = S$ with respect to overlap counting. The imbedded Markov chain $\{Y_t\}$ associated with $W(\Lambda)$ has the transition probability matrix given in Example 5.7, where it was shown that $\Phi_W(s) = (s + ps^2)/(1 - qs - qps^2)$ and $\varphi_W(s) = (p^2 s^2)/(1 - qs - qps^2)$.

Since $\xi(E^o) = (0,1)$, it follows from Example 5.7 and Theorem 5.6 that

$$\phi_1^o(s) = sq(\phi_1^o(s) + \phi_2^o(s)),$$

and that

$$\phi_2^o(s) = s + sp\phi_1^o(s).$$

Solving these simultaneous equations gives

$$\Phi_{W^o}(s) = \frac{s}{1 - qs - qps^2},$$

and

$$\varphi_{W^o}(s) = 1 + \Phi_{W^o}(s) - \frac{1}{s}\Phi_{W^o}(s) = \frac{ps(1 - qs)}{1 - qs - qps^2}.$$

It follows that

$$\begin{aligned} \varphi_{W^o(r,\Lambda)}(s) &= \varphi_W(s)(\varphi_{W^o}(s))^{r-1} \\ &= \frac{p^{r+1}s^{r+1}(1 - qs)^{r-1}}{(1 - qs - qps^2)^r}, \end{aligned}$$

and

$$EW^o(r,\Lambda) = \frac{1}{p} + \frac{r}{p^2} = (\frac{1}{p} + \frac{1}{p^2}) + \frac{r-1}{p^2}.$$

$\diamond$

In general, it follows from Eq. (5.50) that

$$\frac{d}{ds}\varphi_{W^o(r,\Lambda)}(s)|_{s=1} = \Phi_W(1) + (r-1)\Phi_{W^o}(1).$$

This is equivalent to saying that

$$EW^o(r,\Lambda) = EW(\Lambda) + (r-1)EW^o(\Lambda).$$

5.9   Scan Statistics

Let $\{X_i\}$ be a sequence of n *i.i.d.* or homogeneous Markov-dependent two-state trials. The scan statistic $S_n(r)$ of window size r for the sequence $\{X_i\}$ is defined as

$$S_n(r) = \max_{r \le t \le n} S(r,t), \tag{5.51}$$

where $S(r, t)$ is the partial sum given by

$$S(r, t) = \sum_{k=t-r+1}^{t} X_k.$$

The scan statistic has been successfully used in numerous areas, such as cluster analysis (Naus 1965 and Huntington and Naus 1975), the generalized birthday problem (Saperstein 1972 and Naus 1974), DNA pattern matching (Sheng and Naus 1994) and the reliability of k-within-consecutive-r-out-of-n:F systems (Chao, Fu and Koutras 1995).

Naus (1974, Theorem 1) provided a combinatorial result for the conditional distribution of $S_n(r)$ given $\sum_{i=1}^{n} X_i = a$, the total number of successes in n trials. The result is briefly described in the following.

Consider n positions (trials) being divided into L disjoint groups each of size r, *i.e.* $n = rL$. Let σ denote a partition of a 1s into L numbers $(n_1, \cdots, n_L)$, where $n_i \geq 0$ represents the number of 1s in the i-th group and $\sum_{i=1}^{L} n_i = a$. Let F_s denote the collection of all such partitions of a where the L non-negative integers are each smaller than s (*i.e.* $n_i < s$ for $i = 1, \cdots, L$). By using a theorem of Karlin and McGregor (1959), Naus (1974) gave the following formula

$$P\left(S_n(r) < s \middle| \sum_{i=1}^{n} X_i = a\right) = \frac{(r!)^L}{\binom{n}{a}} \sum_{\sigma \in F_s} \det |d_{ij}|, \qquad (5.52)$$

where $d_{ij} = 1/(c_{ij}!(r - c_{ij})!)$, $d_{ij} \equiv 0$ if any of the factorials is less than zero, and, for $i, j = 1, \cdots, L$,

$$c_{ij} = \begin{cases} (j-i)s - \sum_{k=1}^{j-1} n_k + n_i & \text{if } i < j \\ (j-i)s + \sum_{k=j}^{i} n_k & \text{if } i \geq j. \end{cases}$$

The unconditional probability $P(S_n(r) < s)$ can be obtained by averaging the conditional probability $P(S_n(r) < s | \sum_{i=1}^{n} X_i = a)$ over the binomial distribution of $\sum_{i=1}^{n} X_i$, *i.e.*

$$P(S_n(r) < s) = \sum_{a=0}^{n} \binom{n}{a} p^a q^{n-a} P\left(S_n(r) < s \middle| \sum_{i=1}^{n} X_i = a\right).$$

The major difficulty encountered in the computation for large n is that the conditional probability involves a sum of large determinants ($\det |d_{ij}|$) over many items in $\sum_{\sigma \in F_s}$, and the number of terms in this sum tends to infinity, as $n \to \infty$, exponentially fast. Naus (1982) pointed out that Eq. (5.52) can be computed only if r is small and the ratio r/n is relatively large. Even with today's computers, r is still restricted to about less than 20 and n to less than 100 when Eq. (5.52) is used. Because of the complexity of computing the exact probability, a considerable number of approximations and bounds have been developed for $P(S_n(r) < s)$; see, for example, Naus (1982), Glaz (1989, 1992), and Chen and Glaz (1997, 1999). Some of the bounds are excellent. Two recent books, one by Glaz, Naus, and Wallenstein (2001) and the other by Balakrishnan and Koutras (2002), provide comprehensive studies of various types of scan statistics.

Koutras and Alexandrou (1995) studied the exact distribution of the scan statistic $S_n(r)$ using the finite Markov chain imbedding technique. The following alternative formula for the exact distribution is based on Fu (2001), where the tail probability of the scan statistic $P(S_n(r) < s)$ is cast in terms of the transition probability matrix of the imbedded Markov chain corresponding to the waiting time of a compound pattern $\Lambda = \cup_{i=1}^{l} \Lambda_i$. To make this link more clear, we first give the following example.

Example 5.10 Suppose that the window size is $r = 5$ and that $s = 3$, then the event $S_n(5) < 3$ occurs in a sequence of n trials $\{X_i\}_{i=1}^{n}$ if and only if none of the six simple patterns $\Lambda_1 = SSS$, $\Lambda_2 = SFSS$, $\Lambda_3 = SSFS$, $\Lambda_4 = SFFSS$, $\Lambda_5 = SFSFS$, and $\Lambda_6 = SSFFS$ occur in the sequence. If we define the compound pattern $\Lambda_{5,3} = \cup_{i=1}^{6} \Lambda_i$, and let $W(\Lambda_{5,3})$ be the waiting time of the compound pattern $\Lambda_{5,3}$, then we have

$$P(S_n(5) < 3) = P(W(\Lambda_{5,3}) \geq n+1).$$

Furthermore, we may note that the number of simple patterns Λ_i associated with the event $S_n(5) < 3$ is

$$l = \binom{1}{0} + \binom{2}{1} + \binom{3}{2} = 6.$$

From this example, it should be clear that it is sufficient to find the distribution of the waiting time of the compound pattern Λ when we want to find the distribution of $S_n(r)$. $\diamond$

In general, given r and s, where $0 < s \leq r$, we define a collection of simple patterns

$$\mathcal{F}_{r,s} = \{\Lambda_i : \Lambda_1 = \underbrace{S \cdots S}_{s}, \Lambda_2 = SFS \cdots S, \cdots, \Lambda_l = \underbrace{S \cdots SF \cdots FS}_{r}\}$$

$$(5.53)$$

and the compound pattern

$$\Lambda_{r,s} = \cup_{i=1}^{l} \Lambda_i \qquad (5.54)$$

induced by all simple patterns $\Lambda_i \in \mathcal{F}_{r,s}$. We define the usual waiting-time random variable for the compound pattern $\Lambda_{r,s}$ as

$$
\begin{aligned}
W(\Lambda_{r,s}) \quad = \quad & \inf\{n: \text{ the number of trials required to obtain the compound} \\
& \text{pattern } \Lambda_{r,s}\} \\
= \quad & \text{the minimum number of trials required to obtain any one of} \\
& \text{the simple patterns } \Lambda_i, \ \Lambda_i \in \mathcal{F}_{r,s}. \qquad (5.55)
\end{aligned}
$$

Theorem 5.11   *Given r, s, and n, where $0 < s \leq r \leq n$,*

(i) *the waiting-time random variable associated with the scan statistic $S_n(r)$ is $W(\Lambda_{r,s})$, where $\Lambda_{r,s}$ is a compound pattern defined by Eqs. (5.53) and (5.54), and the number of simple pattern is*

$$l = \sum_{x=0}^{r-s} \binom{s-2+x}{x},$$

(ii) *the imbedded homogeneous Markov chain $\{Y_t\}$ is defined on the state space*

$$\Omega = \{\emptyset, S, F\} \cup_{i=1}^{l} \mathbb{S}(\Lambda_i), \quad \text{for } \Lambda_i \in \mathcal{F}_{r,s},$$

and has a transition probability matrix

$$\boldsymbol{M}_{r,s} = \left(\begin{array}{c|c} \boldsymbol{N}_{r,s} & \boldsymbol{C}_{r,s} \\ \hline \boldsymbol{0} & \boldsymbol{I} \end{array} \right),$$

where the transition probabilities $p_{u,v}$ of the matrix $\boldsymbol{M}_{r,s}$ are defined by Eq. (5.14), and

(iii) *we have*

$$P(S_n(r) < s) = P(W(\Lambda_{r,s}) \geq n+1) = \boldsymbol{\xi} \boldsymbol{N}_{r,s}^n \boldsymbol{1}'.$$

Proof. Given r, s and n, it is clear that if $S_n(r) < s$, then no $\Lambda_i \in \mathcal{F}_{r,s}$ has occurred in the sequence $\{X_i\}_{i=1}^n$, and the converse is also true. Hence the waiting time of the compound pattern $\Lambda_{r,s}$ is related to $S_n(r)$ in the following way:

$$S_n(r) < s \Longleftrightarrow W(\Lambda_{r,s}) \geq n+1, \text{ for any } 1 \leq s \leq r.$$

This proves the first parts of Theorem 5.11(i) and (iii). Let x be the number of Fs between the first and the last S (the s-th S). In view of Example 5.10 and the definition of $\mathcal{F}_{r,s}$, the patterns in $\mathcal{F}_{r,s}$ are created by the permutations of $(s-2)$ Ss and x ($x = 0, \cdots, r-s$) Fs between the first and last S, and hence the number of patterns in $\mathcal{F}_{r,s}$ is

$$l = \binom{s-2}{0} + \binom{s-2+1}{1} + \cdots + \binom{s-2+r-s}{r-s} = \sum_{x=0}^{r-s} \binom{s-2+x}{x}.$$

This completes the proof of the second part of Theorem 5.11(i).

Each subpattern in $\mathcal{S}(\Lambda_i)$, $i = 1, \cdots, l$, has to be a state in Ω with respect to the imbedded Markov chain $\{Y_t\}$, and $\emptyset$ is the initial state; hence $\Omega = \{\emptyset, F, S\} \cup_{i=1}^l \mathcal{S}(\Lambda_i)$.

It follows from Section 5.3 that the waiting-time random variable $W(\Lambda_{r,s})$ is finite Markov chain imbeddable. The transition probability matrix $\boldsymbol{M}$ of the imbedded Markov chain $\{Y_t\}$ can be rearranged in the form

$$\boldsymbol{M} = \begin{pmatrix} \boldsymbol{N}_{r,s} & \boldsymbol{C}_{r,s} \\ \hline \boldsymbol{0} & \boldsymbol{I} \end{pmatrix},$$

where the absorbing states $\alpha_1, \cdots, \alpha_l$ correspond to the patterns $\Lambda_i \in \mathcal{F}_{r,s}$. Hence Theorem 5.11(ii) is an immediate consequence of Theorem 5.2. This completes the proof. □

Again we provide an example of the transition probability matrices of the imbedded Markov chain for both *i.i.d.* and Markov-dependent two-state trials.

Example 5.11 Given $r = 4$ and $s = 3$, it follows that $\mathcal{F}_{r,s} = \{\Lambda_1 = SSS, \Lambda_2 = SFSS \text{ and } \Lambda_3 = SSFS\}$ and that the state space is $\Omega = \{\emptyset, F, S, SF, SS, SFS, SSF, \alpha\}$ after combining the absorbing states α_1, α_2 and α_3 together as α. If $\{X_i\}$ is a sequence of *i.i.d.* two-state trials, then the transition probability matrix of the imbedded Markov chain $\{Y_t\}$ defined on Ω has the form

$$
M = \begin{matrix} \emptyset \\ F \\ S \\ SF \\ SS \\ SFS \\ SSF \\ \alpha \end{matrix}
\left(\begin{array}{ccccccc:c}
0 & q & p & 0 & 0 & 0 & 0 & 0 \\
0 & q & p & 0 & 0 & 0 & 0 & 0 \\
0 & 0 & 0 & q & p & 0 & 0 & 0 \\
0 & q & 0 & 0 & 0 & p & 0 & 0 \\
0 & 0 & 0 & 0 & 0 & 0 & q & p \\
0 & 0 & 0 & q & 0 & 0 & 0 & p \\
0 & q & 0 & 0 & 0 & 0 & 0 & p \\
\hdashline
0 & 0 & 0 & 0 & 0 & 0 & 0 & 1
\end{array}\right)
= \left(\begin{array}{c:c} N_{4,3} & C_{4,3} \\ \hdashline 0 & I \end{array}\right).
$$

If the sequence of two-state trials is a homogeneous Markov chain with transition probability matrix

$$
A = \left(\begin{array}{cc} p_{FF} & p_{FS} \\ p_{SF} & p_{SS} \end{array}\right),
$$

then the transition probability matrix M^* corresponding to the imbedded Markov chain $\{Y_t\}$ is

$$
M^* = \begin{matrix} \emptyset \\ F \\ S \\ SF \\ SS \\ SFS \\ SSF \\ \alpha \end{matrix}
\left(\begin{array}{ccccccc:c}
0 & q & p & 0 & 0 & 0 & 0 & 0 \\
0 & p_{FF} & p_{FS} & 0 & 0 & 0 & 0 & 0 \\
0 & 0 & 0 & p_{SF} & p_{SS} & 0 & 0 & 0 \\
0 & p_{FF} & 0 & 0 & 0 & p_{FS} & 0 & 0 \\
0 & 0 & 0 & 0 & 0 & 0 & p_{SF} & p_{SS} \\
0 & 0 & 0 & p_{SF} & 0 & 0 & 0 & p_{SS} \\
0 & p_{FF} & 0 & 0 & 0 & 0 & 0 & p_{FS} \\
\hdashline
0 & 0 & 0 & 0 & 0 & 0 & 0 & 1
\end{array}\right).
$$

Note that here we assume $P(X_1 = S) = p$, $P(X_1 = F) = q$, and the initial distribution $P(Y_0 = \emptyset) = 1$ in both cases. The two transition probability matrices share the same form, and the changes from matrix M to matrix M^* are straightforward. Since $P(W(\Lambda_{r,s}) \geq n+1) = P(S_n(r) < s)$, application of Theorem 5.6 yields the probability generating function for the sequence $\{P(S_n(4) < 3)\}$ as follows:

$$
\Phi_W(t) = \frac{t(1 + pt + p^2 t^2 + p^2 q t^3 - p^3 q t^4 - p^3 q^2 t^5)}{1 - qt - pqt^2 - p^2 q^2 t^4 + p^3 q^3 t^6}.
$$

$\diamond$

Tables 5.2 and 5.3 provide selected numerical examples. Specifically, Table 5.2 provides some cumulative distributions of the scan statistic $S_n(r)$

for $r = 5$ and $n = 10, 50, 500$, and Table 5.3 provides selected probabilities $P(S_n(r) < s)$ for $r = 10, 15, 20$ and large n.

Table 5.2   Cumulative distribution of $S_n(5)$.

Selected case		*i.i.d.*	Markov-dependent
n	s	$p = q = 0.5$	$p_{FF} = 0.3, p_{SF} = 0.4$
10	0	0.0009766	0.0050388
	1	0.0253906	0.0965779
	2	0.1816406	0.4486013
	3	0.5478516	0.8453266
	4	0.8906250	0.9860426
	5	1	1
Mean		3.3535156	2.6184271
Standard deviation		0.9581207	0.9024778
50	0	8.881784×10^{-16}	$6.7356773 \times 10^{-12}$
	1	1.7635697×10^{-9}	2.9342188×10^{-6}
	2	0.0000418	0.0085945
	3	0.0215953	0.3396303
	4	0.4481247	0.9071361
	5	1	1
Mean		3.9830427	3.1873302
Standard deviation		0.7501573	0.7636984
500	0	$3.0549364 \times 10^{-151}$	$9.9184531 \times 10^{-112}$
	1	$5.4756424 \times 10^{-90}$	$4.6027994 \times 10^{-57}$
	2	$5.0166948 \times 10^{-46}$	$4.1206202 \times 10^{-22}$
	3	$3.4424189 \times 10^{-18}$	1.1910137×10^{-5}
	4	0.0001974	0.3549406
	5	1	1
Mean		4.5302382	3.7446362
Standard deviation		0.5407803	0.6269622

As can be seen from Tables 5.2 and 5.3, given a value of r, for every $s \leq r$ the probability $P(S_n(r) < s)$ tends to zero very fast as n increases. In fact, even $P(S_n(r) < r)$ tends to zero exponentially fast in the sense that

$$\lim_{n \to \infty} \frac{1}{n} \log\{P(S_n(r) < r)\} = -\beta, \qquad (5.56)$$

Table 5.3   Selected $P(S_n(r) < s)$ for large r and n.

Selected case			i.i.d.	Markov-dependent
n	r	s	$p = q = 0.5$	$p_{FF} = 0.3, p_{SF} = 0.4$
100	15	4	$6.7451941 \times 10^{-15}$	2.1329403×10^{-9}
	20	4	$6.6456841 \times 10^{-18}$	$4.1583113 \times 10^{-11}$
500	10	3	$1.4369406 \times 10^{-79}$	$6.2128156 \times 10^{-48}$
	10	5	$4.0460520 \times 10^{-34}$	$7.4002280 \times 10^{-14}$
	15	4	$1.1297610 \times 10^{-73}$	$2.2560900 \times 10^{-43}$
	20	4	$1.1020938 \times 10^{-86}$	$2.3334601 \times 10^{-53}$

where $\beta = -\log \lambda_{[1]}$ and $\lambda_{[1]}$ is the largest eigenvalue of the essential transition probability submatrix of the imbedded Markov chain $\{Y_t\}$ corresponding to the waiting-time random variable $W(\Lambda_{r,r})$. This is a direct result of Theorem 5.9.

In this section, we have focused our study on the scan statistic $S_n(r)$. There are two other commonly used scan statistics, $W_{s,r}$ and $D_n(s)$, as defined below:

(i) Given s and r, let $W_{s,r}$ denote the waiting time until the first occurrence of s successes within a window of length r,

$$W_{s,r} = \inf\{t : S(r, t) \geq s\}.$$

(ii) Given n and s, let $D_n(s)$ be the length of the smallest window that contains at least s successes,

$$D_n(s) = \inf\{r : S_n(r) \geq s\}.$$

From the definitions, the two statistics $W_{s,r}$ and $D_n(s)$ are closely related to the scan statistic $S_n(r)$, and their distributions can be obtained in terms of the distribution of $S_n(r)$. These relationships can be mathematically expressed as

$$P(S_n(r) < s) = P(W_{r,s} > n) = P(D_n(s) > r), \quad \text{for } 0 < s \leq r \leq n. \quad (5.57)$$

Chapter 6

Random Permutations

6.1 Introduction

Let $Z_n = \{1, 2, \cdots, n\}$ be a collection of n integers, and let $\Pi(Z_n) = \{\pi_n = (\pi(1), \pi(2), \cdots, \pi(n)) : \pi(i) \in Z_n\}$ be the set of all permutations generated by the integers in Z_n. A *succession of size 2* has occurred at the i-th place if $\pi(i) + 1 = \pi(i+1)$, and a *rise* (increase) has occurred at the i-th place if $\pi(i) < \pi(i+1)$. Successions and rises can be viewed as patterns in the permutation π_n. For example, the number of permutations in $\Pi(Z_n)$ with exactly k rises, $A(n, k)$, is known as the Eulerian number.

More generally, let S be a collection of N integers with $[s]$-specification $[s] = [s_1, \cdots, s_n]$, where $s_i (\geq 1)$, for $i = 1, \cdots, n$, is the number of integers i and $s_1 + \cdots + s_n = N$. For example, if $S = \{1, 1, 2, 3, 3\}$, then $[s] = [2, 1, 2]$ and $s_1 + s_2 + s_3 = 5$. Denote by $\Pi_N(S) = \{\pi_N = (\pi(1), \cdots, \pi(N)) : \pi(i) \in S\}$ the collection of all the permutations generated by the integers in S. The number of permutations in $\Pi_N(S)$ having exactly k rises (or falls), $A([s], k)$, is called the Simon Newcomb number. It is easy to see that $\Pi(Z_n)$ and $A(n, k)$ are special cases of $\Pi_N(S)$ and $A([s], k)$, respectively, when the $[s]$-specification is $[1, \cdots, 1]$.

In the literature, the Eulerian and Simon Newcomb numbers are probably the most heavily studied and celebrated numbers associated with random permutations in combinatorial analysis. There are many other interesting patterns and runs statistics besides these two, such as the numbers of *levels*, *picks* and *waves*. A considerable amount of literature in combinatorial analysis has treated such numbers of runs and patterns in a random permutation $\pi_N \in \Pi_N(S)$: *e.g.* Abramson and Moser (1967), Dwass (1973),

Jackson and Reilly (1976), Johnson (2001), Kaplansky (1944), Reilly and Tanny (1979), Roselle (1968) and Tanny (1976). The books by MacMahon (1915), Riordan (1958) and David and Barton (1962) provide a good overview of early developments in this area; Johnson (2002) gave a short review of current developments. Recently Fu (1995), Fu, Lou and Wang (1999) and Johnson and Fu (2000) extended the finite Markov chain imbedding technique to study the distribution of runs and patterns defined on random permutations. Whenever the context is clear, we will simplify the notation π_n and π_N to π.

If we assume that all the permutations in $\Pi(Z_n)$ are equally likely, then, for example, the probability of exactly k rises in a permutation π can be written as

$$P(R_n(\pi) = k) = A(n, k)/n!,$$

where $R_n(\pi)$ denotes the number of rises in a random permutation π. Furthermore, note that every permutation $\pi \in \Pi(Z_n)$ is a realization of inserting n integers one by one into gaps between integers (including the two end gaps). For example, consider first that the permutation $\pi = (31524687)$ can be decomposed into 8 sub-permutations

$$\{\pi_8 = \pi, \pi_7 = (3152467), \pi_6 = (315246), \pi_5 = (31524),$$
$$\pi_4 = (3124), \pi_3 = (312), \pi_2 = (12), \text{ and } \pi_1 = (1)\}$$

by removing the largest integer at each step. This decomposition is unique. Conversely, the random permutation π_8 is a realization of inserting the integers 1 to 8 one by one randomly into the gaps between integers:

$$\{(1) \rightarrow (12) \rightarrow (312) \rightarrow \cdots \rightarrow (3152467) \rightarrow (31524687)\}.$$

In general, under the above random insertion procedure, all the permutations in $\Pi(Z_n)$ are equally likely and have probability $1/n!$. The most important aspect of the above insertion procedure is that it provides a way to study the distributions of runs and patterns on random permutations via the finite Markov chain imbedding technique without combinatorial analysis. Basically, the sequence of insertions will be viewed as a sequence of trials, and treated as in previous chapters.

6.2 Successions

Let $\pi = (\pi(1), \cdots, \pi(n))$ be a random permutation in $\Pi(Z_n)$. A *succession of size 2* is a pair of integers $(\pi(i), \pi(i+1))$ which satisfy

$$\pi(i+1) = \pi(i) + 1. \tag{6.1}$$

The number of successions of size 2 in π, $X_n(\pi, 2)$, is the total number of pairs $(\pi(i), \pi(i+1))$, $i = 1, 2, \cdots, (n-1)$, that satisfy the condition (6.1) (with overlap counting). More generally, for $2 \leq m \leq n$, we define a sequence of index functions for successions of size m: for $i = 1, \cdots, n-m+1$,

$$I_n(\pi, m, i) = \begin{cases} 1 & \text{if } \pi(i+m-1) = \pi(i+m-2) + 1 \\ & = \cdots = \pi(i) + m - 1 \\ 0 & \text{otherwise.} \end{cases} \tag{6.2}$$

We then define the random variable

$$X_n(\pi, m) = \sum_{i=1}^{n-m+1} I_n(\pi, m, i), \tag{6.3}$$

the total number of successions of size m in the permutation π.

Every permutation $\pi \in \Pi(Z_n)$ can be viewed as a sequence of n sub-permutations $\{\pi_1, \pi_2, \cdots, \pi_n\}$, where π_t is obtained by inserting the integer t into one of the gaps of the sub-permutation π_{t-1}, for $t = 2, \cdots, n$, with $\pi_1 = (1)$. Let $\Omega = \{0, 1, 2, \cdots, n-1\}$ be a state space and $\Gamma_n = \{0, 1, \cdots, n\}$ be an index set. For $m = 2$, we define a sequence of transformations $Y_t : \Pi(Z_n) \to \Omega$, for $t = 1, 2, \cdots, n$ and every $\pi \in \Pi(Z_n)$, as

$$
\begin{aligned}
Y_t(\pi) &= X_t(\pi_t, 2) \\
&= \text{the total number of successions of size 2 in the} \\
&\quad \text{sub-permutation } \pi_t \text{ induced by } \pi.
\end{aligned} \tag{6.4}
$$

For example, consider again the permutation $\pi = (31524687)$, which can be decomposed into 8 sub-permutations by deleting the largest integer one by one from the permutation π: $\pi_8 = \pi$, $\pi_7 = (3152467)$, $\pi_6 = (315246)$, $\pi_5 = (31524)$, $\pi_4 = (3124)$, $\pi_3 = (312)$, $\pi_2 = (12)$ and $\pi_1 = (1)$. Hence π is a realization of the random insertion procedure. Further, the realization of the imbedded chain $\{Y_t(\pi)\}_{t=1}^8$ with respect to $\pi = (31524687)$ is $\{Y_1(\pi) = 0, Y_2(\pi) = 1, Y_3(\pi) = 1, Y_4(\pi) = 1, Y_5(\pi) = 0, Y_6(\pi) = 0, Y_7(\pi) = 1$ and $Y_8(\pi) = 0\}$. From the method of insertion and the above example, it may be observed that if $Y_{t-1}(\pi) = k$ (for $t = 2, \cdots, n$ and $0 \leq k \leq t - 2$), then $Y_t(\pi)$ can only be at state $k - 1, k$ or $k + 1$. Since the integer "t" has equal probability of being inserted into any one of t positions (gaps) of the sub-permutation π_{t-1}, the transition probabilities associated with the imbedded chain $\{Y_t\}$ are specified by the following equation: for $t = 2, \cdots, n$ and $0 \leq k \leq t - 2$,

$$P(Y_t = x | Y_{t-1} = k) = \begin{cases} k/t & \text{if } x = k - 1 \\ (t - k - 1)/t & \text{if } x = k \\ 1/t & \text{if } x = k + 1, \end{cases} \tag{6.5}$$

with $P(Y_0 = 0) = 1$ and $P(Y_1 = 0 | Y_0 = 0) = 1$. It follows from Eqs. (6.4) and (6.5) that the sequence $\{Y_t : t \in \Gamma_n\}$ forms a non-homogeneous Markov chain on Ω having transition probability matrices given by

$$M_t(n) = \begin{matrix} 0 \\ 1 \\ 2 \\ \vdots \\ \vdots \\ t-3 \\ t-2 \\ t-1 \\ \vdots \\ n-1 \end{matrix} \left(\begin{matrix} \frac{t-1}{t} & \frac{1}{t} & & & & & & & \\ \frac{1}{t} & \frac{t-2}{t} & \frac{1}{t} & & & O & & & O \\ & \frac{2}{t} & \frac{t-3}{t} & \frac{1}{t} & & & & & \\ & & \ddots & \ddots & \ddots & & & & \\ & & & \ddots & \ddots & \ddots & & & \\ & O & & & \ddots & \ddots & \ddots & & \\ & & & & & \frac{t-3}{t} & \frac{2}{t} & \frac{1}{t} & \\ & & & & & & \frac{t-2}{t} & \frac{1}{t} & \frac{1}{t} \\ \hline & & & O & & & & & I_{n-t+1} \end{matrix} \right), \tag{6.6}$$

for $t = 2, \cdots, n$, where $M_t(n)$ contains the entries $p_{ij}(t : n)$,

$$p_{ij}(t : n) = P(Y_t = j | Y_{t-1} = i), \quad i, j = 0, 1, \cdots, t - 2,$$

defined by Eq. (6.5), and I_{n-t+1} is the $(n - t + 1) \times (n - t + 1)$ identity matrix.

Let $\boldsymbol{a}(t) = (a_0(t), a_1(t), \cdots, a_{n-1}(t))$, $t = 1, 2, \cdots, n$, be n vectors with components

$$a_i(t) = P(Y_t = i | Y_0 = 0), \tag{6.7}$$

for $i = 0, 1, \cdots, n - 1$. For $t = 0$, the initial distribution of Y_0 gives $\boldsymbol{a}(0) = (1, 0, \cdots, 0)$, and for $t = 1$, the transition probability matrix $M_1(n)$ has entry $p_{00}(1 : n) = P(Y_1 = 0|Y_0 = 0) = 1$ and has zero at all the other entries.

It follows from Eqs. (6.4), (6.5), and (6.6) that $X_n(\pi, 2)$ is finite Markov chain imbeddable with respect to the insertion procedure, and that

$$P(X_n(\pi, 2) = i) = P(Y_n(\pi) = i|Y_0 = 0). \qquad (6.8)$$

Theorem 6.1 *Given the initial distribution $\boldsymbol{a}(0)$, the distribution of $X_n(\pi, 2)$ can be obtained by*

$$P(X_n(\pi, 2) = i) = \boldsymbol{a}(0) \left(\prod_{t=1}^{n} \boldsymbol{M}_t(n) \right) \boldsymbol{e}_i', \quad i = 0, 1, \cdots, n-1, \qquad (6.9)$$

where $\boldsymbol{M}_t(n)$ is given by Eq. (6.6). Further, for $n > 1$, $a_i(n)$ satisfies the following recursive equation:

$$a_i(n) = \begin{cases} \frac{n-1}{n}a_0(n-1) + \frac{1}{n}a_1(n-1) & \text{if } i = 0 \\ \frac{1}{n}a_{i-1}(n-1) + \frac{n-i-1}{n}a_i(n-1) \\ \quad + \frac{i+1}{n}a_{i+1}(n-1) & \text{if } 1 \le i \le n-3 \\ \frac{1}{n}a_{n-3}(n-1) + \frac{1}{n}a_{n-2}(n-1) & \text{if } i = n-2 \\ \frac{1}{n}a_{n-2}(n-1) & \text{if } i = n-1. \end{cases} \qquad (6.10)$$

Proof. Since $X_n(\pi, 2)$ is finite Markov chain imbeddable and $\{Y_t\}$ is a non-homogeneous Markov chain with transition probability matrices given by Eq. (6.6), the first part of the theorem follows immediately from Theorem 2.1. The recursive equation follows from

$$\begin{aligned} a_i(n) &= \boldsymbol{a}(0) \left(\prod_{t=1}^{n} \boldsymbol{M}_t(n) \right) \boldsymbol{e}_i' \\ &= (a_0(n-1), a_1(n-1), \cdots, a_{n-1}(n-1)) \boldsymbol{M}_n(n) \boldsymbol{e}_i'. \end{aligned}$$ $\square$

The recursive equation (6.10) provides a simple and direct way to show that the limiting distribution of the number of successions of size 2, $X_n(\pi, 2)$, is a Poisson distribution with parameter $\lambda = 1$.

Theorem 6.2 *Given $\boldsymbol{a}(0) = (1, 0, \cdots, 0)$,*

$$\lim_{n \to \infty} P(X_n(\pi, 2) = x) = \frac{1}{x!}e^{-1}, \quad x = 0, 1, \cdots. \qquad (6.11)$$

In order to prove the above result, we need the following simple lemma.

Lemma 6.1 *Given $\boldsymbol{a}(0) = (1, 0, \cdots, 0)$, it follows that* *(i)* $a_x(x+1) = [(x+1)!]^{-1}$, *for* $x = 0, 1, 2, \cdots, n-1$,

(ii) $a_x(t) = 0$, *for* $t \leq x$.

Proof. (i) For $Y_{x+1} = x$, the permutation π_{x+1} has to have the form $(123 \cdots x(x+1))$, and this event has probability $1/(x+1)!$. (ii) For $t \leq x$, $Y_t = x$ cannot happen, and hence $a_x(t) \equiv 0$. $\qquad\qquad \square$

Proof. (Theorem 6.2). For $x = 0$, it follows from Eq. (6.10) and some simple algebra that

$$
\begin{aligned}
a_0(n) &= (1 - \frac{1}{n})a_0(n-1) + \frac{1}{n}a_1(n-1) \\
&= \cdots \\
&= (1 - \frac{1}{n})^{n-1}[a_0(1) + o(1)].
\end{aligned}
\tag{6.12}
$$

Note that $a_0(1) = 1$. Furthermore, for fixed x $(x \geq 1)$, it again follows from Eq. (6.10) that

$$
\begin{aligned}
a_x(n) &= \frac{1}{n}a_{x-1}(n-1) + \frac{n-x-1}{n}a_x(n-1) + \frac{x+1}{n}a_{x+1}(n-1) \\
&= \cdots \\
&= (1 - \frac{1}{n})^{n-x-1}[a_{x-1}(x) + o(1)].
\end{aligned}
\tag{6.13}
$$

Taking $n \to \infty$, the result (6.11) follows immediately from Eqs. (6.12) and (6.13), and from Lemma 6.1. This completes the proof. $\qquad\qquad \square$

For $m \geq 3$, the random variable $X_n(\pi, m)$ has a degenerate limiting distribution at zero.

Theorem 6.3 *For $m \geq 3$,*

$$
\lim_{n \to \infty} P(X_n(\pi, m) = x) = \begin{cases} 1 & \text{if } x = 0 \\ 0 & \text{if } x \geq 1. \end{cases}
\tag{6.14}
$$

Proof. Let, for $i = 1, 2, \cdots, n - m + 1$, E_i be the event such that $\pi(i) + m - 1 = \pi(i+1) + m - 2 = \cdots = \pi(i+m-1)$. It follows from the definition of E_i and the random insertion procedure that $P(E_i) = 1/n^{m-1} + o(1/n^{m-1})$. From the Bonferroni inequality, it follows that

$$
P(X_n(\pi, m) \geq 1) = P(\cup_{i=1}^{n-m+1} E_i) \leq \sum_{i=1}^{n-m+1} P(E_i)
\tag{6.15}
$$

$$= \sum_{i=1}^{n-m+1} \left[\frac{1}{n^{m-1}} + o\left(\frac{1}{n^{m-1}}\right)\right]$$

$$= \frac{1}{n^{m-2}} + o\left(\frac{1}{n^{m-2}}\right).$$

Taking $n \to \infty$, the result (6.14) is an immediate consequence of inequality (6.15). $\qquad\square$

From Theorems 6.2 and 6.3, we may conclude that in a very large random permutation, the number of successions of size 2 has a Poisson distribution with $\lambda = 1$, and that, with probability tending to one, there are no successions of size greater than 2.

To illustrate Theorems 6.1 and 6.2, we give in Table 6.1 some numerical results for the exact distribution of $X_n(\pi, 2)$ and its limiting distribution, the Poisson distribution with $\lambda = 1$.

Since the number of insertions can be viewed as an index set, we can define a waiting-time (number of insertions) random variable as follows:

$$W(k;2) = \text{the smallest number of insertions required to have}$$
$$k \text{ successions of size 2.} \qquad (6.16)$$

With a minor modification to the Markov chain $\{Y_t\}$ defined by Eq. (6.4), we are able to obtain a new imbedded Markov chain for the waiting-time random variable $W(k;2)$. Given k, we define a Markov chain $\{Y_t(k); t = 1, 2, \cdots\}$ on the state space $\Omega_k = \{0, 1, \cdots, k-1, k\}$ with the following transition probability matrices: for $2 \le t \le k+1$,

$$\boldsymbol{M}_t(k) = \begin{array}{c} 0 \\ 1 \\ 2 \\ \vdots \\ \vdots \\ t-3 \\ t-2 \\ t-1 \\ \vdots \\ k \end{array} \left(\begin{array}{cccccc|c} \frac{t-1}{t} & \frac{1}{t} & & & & & \\ \frac{1}{t} & \frac{t-2}{t} & \frac{1}{t} & & \boldsymbol{O} & & \\ & \frac{2}{t} & \frac{t-3}{t} & \frac{1}{t} & & & \\ & & \ddots & \ddots & \ddots & & \boldsymbol{O} \\ \boldsymbol{O} & & & \ddots & \ddots & \ddots & \\ & & & & \frac{t-3}{t} & \frac{2}{t} & \frac{1}{t} \\ & & & & & \frac{t-2}{t} & \frac{1}{t} \mid \frac{1}{t} \\ \hline & & \boldsymbol{O} & & & & \boldsymbol{I}_{k-t+2} \end{array} \right),$$

Table 6.1 Selected distributions of $X_n(\pi, 2)$, and its limiting distribution, the Poisson distribution with $\lambda = 1$.

$x \backslash n$	5	10	15	20	Poisson(1)
0	.44167	.40467	.39241	.38627	.36788
1	.36667	.36788	.36788	.36788	.36788
2	.15000	.16555	.17168	.17474	.18394
3	.03333	.04905	.05314	.05518	.06131
4	.00833	.01073	.01226	.01303	.01533
5		.00184	.00225	.00245	.00307
6		.00025	.00034	.00038	.00051
7		.00003	.00004	.00005	.00007
8		$.248 \times 10^{-5}$	$.487 \times 10^{-5}$	$.593 \times 10^{-5}$	$.912 \times 10^{-5}$
9		$.276 \times 10^{-6}$	$.473 \times 10^{-6}$	$.608 \times 10^{-6}$	$.101 \times 10^{-5}$
10			$.406 \times 10^{-7}$	$.558 \times 10^{-7}$	$.101 \times 10^{-6}$
11			$.306 \times 10^{-8}$	$.461 \times 10^{-8}$	$.922 \times 10^{-8}$
12			$.209 \times 10^{-9}$	$.346 \times 10^{-9}$	$.768 \times 10^{-9}$
13			$.107 \times 10^{-10}$	$.236 \times 10^{-10}$	$.591 \times 10^{-10}$
14			$.765 \times 10^{-12}$	$.148 \times 10^{-11}$	$.422 \times 10^{-11}$
15				$.844 \times 10^{-13}$	$.281 \times 10^{-12}$
16				$.438 \times 10^{-14}$	$.176 \times 10^{-13}$
17				$.211 \times 10^{-15}$	$.103 \times 10^{-14}$
18				$.781 \times 10^{-17}$	$.575 \times 10^{-16}$
19				$.411 \times 10^{-18}$	$.302 \times 10^{-17}$

and for $t \geq k + 2$,

$$
\boldsymbol{M}_t(k) = \begin{matrix} 0 \\ 1 \\ 2 \\ \vdots \\ \vdots \\ k-1 \\ k \end{matrix}
\begin{pmatrix}
\frac{t-1}{t} & \frac{1}{t} & 0 & & & & \\
\frac{1}{t} & \frac{t-2}{t} & \frac{1}{t} & & \boldsymbol{O} & & \\
0 & \frac{2}{t} & \frac{t-3}{t} & \frac{1}{t} & & & \\
& & \ddots & \ddots & \ddots & & \\
& & & \ddots & \ddots & \ddots & \\
& \boldsymbol{O} & & & \frac{k-1}{t} & \frac{t-k}{t} & \frac{1}{t} \\
& & & & 0 & 0 & 1
\end{pmatrix} . \quad (6.17)
$$

The transition probability matrices $\boldsymbol{M}_t(n)$ and $\boldsymbol{M}_t(k)$ for Markov chains $\{Y_t\}$ and $\{Y_t(k)\}$, respectively, are very similar. The major difference is that the absorbing state "k" of the Markov chain $\{Y_t(k)\}$ corresponds to the states k, $k+1, \cdots$, $(n-1)$ of the Markov chain $\{Y_t\}$. The Markov chain $\{Y_t(k)\}$ specified by the transition probability matrices $M_t(k)$ has two important characteristics:

(i) if $Y_t(k) \le k-1$, it implies $Y_i(k) \le k-1$ for all $i \le t$, and
(ii) if $Y_t(k) = k$, it implies $Y_i(k) = k$ for all $i \ge t$.

These two characteristics and the definitions of the two random variables $Y_t(k)$ and $W(k; 2)$ yield the conclusion that they are one-to-one related in the following way:

$$W(k; 2) \ge n \iff Y_{n-1}(k) \le k-1. \tag{6.18}$$

Hence the following theorem holds.

Theorem 6.4 *For $n \ge k+1$ and $k = 1, 2, \cdots$, we have*

$$P(W(k; 2) = n) = a(0) \left(\prod_{t=1}^{n-1} M_t(k) \right) (I - M_n(k)) U'(k),$$

where $M_t(k)$ is defined by Eq. (6.17), $a(0) = (1, 0, \cdots, 0)_{1 \times (k+1)}$, and $U(k) = (1, 1, \cdots, 1, 0)_{1 \times (k+1)}$.

Proof. It follows from Eq. (6.18) that

$$
\begin{aligned}
P(W(k; 2) = n) &= P(W(k; 2) \ge n) - P(W(k; 2) \ge n+1) \\
&= P(Y_{n-1}(k) \le k-1) - P(Y_n(k) \le k-1) \\
&= a(0) \left(\prod_{t=1}^{n-1} M_t(k) \right) (I - M_n(k)) U'(k).
\end{aligned}
$$

□

For the special case $k = 1$, the distribution of the waiting time for the first succession of size 2 can be determined by the following recursive equation.

Theorem 6.5 *For $n \ge 2$ and the condition $P(W(1; 2) = 1) = 0$,*

$$P(W(1; 2) = n) = \frac{1}{n} \left(1 - \sum_{i=1}^{n-1} P(W(1; 2) = i) \right).$$

Proof. For $k = 1$ and $t = 1$, the transition probability matrix $M_1(1)$ is given by

$$M_1(1) = \begin{pmatrix} 1 & 0 \\ 0 & 0 \end{pmatrix}.$$

For $k = 1$ and $t = 2, 3, \cdots$, the matrices $M_t(k)$ given by Eq. (6.17) reduce to

$$M_t(1) = \begin{matrix} 0 \\ 1 \end{matrix} \begin{pmatrix} \frac{t-1}{t} & \frac{1}{t} \\ 0 & 1 \end{pmatrix}.$$

Since, by Theorem 6.4,

$$
\begin{aligned}
\sum_{i=1}^{n-1} P(W(1;2) = i) &= \sum_{i=1}^{n-1}(1,0)\left(\prod_{t=1}^{i-1} M_t(1)\right)(I - M_i(1))(1,0)' \\
&= 1 - (1,0)\left(\prod_{t=1}^{n-1} M_t(1)\right)(1,0)',
\end{aligned}
$$

we have

$$
\begin{aligned}
P(W(1;2) = n) &= (1,0)\left(\prod_{t=1}^{n-1} M_t(1)\right)(I - M_n(1))(1,0)' \\
&= \frac{1}{n}(1,0)\left(\prod_{t=1}^{n-1} M_t(1)\right)(1,0)' \\
&= \frac{1}{n}\left(1 - \sum_{i=1}^{n-1} P(W(1;2) = i)\right).
\end{aligned}
$$

$\square$

For example, it follows from the above theorem that $P(W(1;2) = 2) = 1/2$, $P(W(1;2) = 3) = 1/6$, and $P(W(1;2) = 4) = 1/12$. Note that (3421) and (2134) are the only two permutations with $W(1;2) = 4$ among the 24 possible permutations obtained from inserting the four integers $1, 2, 3$ and 4. Comparing the result of Theorem 6.4 with Eqs. (5.2) and (5.22), loosely speaking, we could say that the waiting time $W(k;2)$ has a general geometric distribution with respect to the insertion procedure on random permutations.

6.3 Eulerian and Simon Newcomb Numbers

Let us consider the following identity:

$$\frac{1-x}{1-x\exp\{\lambda(1-x)\}} = 1 + \sum_{n=1}^{\infty}\sum_{k=1}^{n} A(n,k)x^k\frac{\lambda^n}{n!}. \qquad (6.19)$$

The integers $A(n,k)$, $k = 1, 2, \cdots, n$, on the right-hand side of Eq. (6.19) are known as the Eulerian numbers. Euler first introduced this identity in his famous book *"Institutiones calculi differentialis"* (1755, pp. 485–487), and the formula for $A(n,k)$ given by him has the following form:

$$A(n,k) = \sum_{j=0}^{k}(-1)^j(k-j)^n\binom{n+1}{j}, \quad n \geq k \geq 1. \qquad (6.20)$$

He also showed that $A(n,k)$ satisfies the recursive equation

$$A(n,k) = kA(n-1,k) + (n-k+1)A(n-1,k-1). \qquad (6.21)$$

Worpitzky (1883) proved that Eulerian numbers may also be defined via the identify

$$x^n = \sum_{k=1}^{n}\binom{x+k-1}{n}A(n,k). \qquad (6.22)$$

Roselle (1968) showed that the Eulerian numbers $A(n,k)$ equal the number of permutations in $\Pi(Z_n)$ with exactly k rises, as mentioned in Section 6.1.

Consider a deck S of N cards containing s_i cards of face value i, for $i = 1, 2, \cdots, n$, with $s_1 + \cdots + s_n = N$. Turn over the top card and put it down face up. Turn over the second card and place it on top of the first card if the face value is less than or equal to the first; otherwise, start a new pile. The Simon Newcomb number $A([s],k)$ with respect to the specification $[s] = [s_1, \cdots, s_n]$ is defined to be the number of all possible permutations of the deck S which result in exactly k piles or rises, a definition equivalent to that given in Section 6.1. Dillon and Roselle (1969) gave the following formula for the Simon Newcomb numbers: for $k = 1, 2, \cdots, n$,

$$A([s],k) = \sum_{j=0}^{k}(-1)^j\binom{N+1}{j}\prod_{i=1}^{n}\binom{s_i+k-j-1}{s_i}. \qquad (6.23)$$

Given a permutation $\pi \in \Pi_N(S)$, we define a random variable $R_N(\pi)$, the number of rises (increases) in permutation π, as

$$R_N(\pi) = \sum_{i=1}^{N} I(\pi, i),$$

where $I(\pi, i)$ are index functions defined as

$$I(\pi, i) = \begin{cases} 1 & \text{if } \pi(i) > \pi(i-1) \\ 0 & \text{otherwise,} \end{cases}$$

for $i = 2, 3, \cdots, N$, and $I(\pi, 1) \equiv 1$ by convention. The insertion procedure associated with the $[s]$-specification considered here can be described as inserting integers one by one randomly between gaps, starting with s_1 integers "1", followed by s_2 integers "2", and continuing this procedure until all the integers have been inserted. The main purpose here is to show that the Simon Newcomb number $A([s], k)$ (or the number of rises $R_N(\pi)$) is finite Markov chain imbeddable with respect to the above insertion procedure. Our approach also establishes a recursive equation for $A([s], k)$. The Eulerian numbers $A(n, k)$ will be treated as a special case of the Simon Newcomb numbers with $[s] = [1, \cdots, 1]$.

Let $s^* = \max(s_1, \cdots, s_n)$ and $N^* = N - s^* + 1$. Dillon and Roselle (1969) showed that N^* is the maximum number of rises (by convention, the first gap is a rise and the last gap is a fall) in a random permutation $\pi \in \Pi_N(S)$. For every $1 \le t \le N$, there exists an integer $h = 0, 1, 2, \cdots, n-1$, such that

$$\sum_{i=1}^{h} s_i < t \le \sum_{i=1}^{h+1} s_i,$$

(with convention $\sum_{i=1}^{0} \equiv 0$). Further, for every t, we define

$$
\begin{aligned}
s_{h+1}(t) &= t - \sum_{i=1}^{h} s_i, \\
l(t) &= t - \sum_{i=1}^{h} s_i - 1 = s_{h+1}(t) - 1, \\
s^*(t) &= \max(s_1, \cdots, s_h, s_{h+1}(t)), \\
N_t^* &= t - s^*(t) + 1, \quad N^* = N - s^*(N) + 1, \\
[s]_t &= [s_1, \cdots, s_h, s_{h+1}(t)],
\end{aligned}
$$

and π_t as the permutation generated by the insertion procedure for the first t integers.

Consider a sequence of transformations $\{Y_t\}_{t=1}^{N} : \Pi_N(\pi) \rightarrow \Omega = \{1, 2, \cdots, N^*\}$ defined by

$$Y_t(\pi) = \text{the total number of rises in } \pi_t, \ t = 1, \cdots, N.$$

Lemma 6.2 *For every t, $1 \leq t \leq N$,*

> *(i) N_t^* is the maximum number of rises in a random permutation π_t generated by integers with specification $[s]_t$, and*
>
> *(ii) $l(t)$ equals the total number of falls and levels containing the integers "$h+1$" in a random permutation π_{t-1} generated by the integers with specification $[s_1, \cdots, s_{h+1}(t) - 1]$.*

Proof. Result (i) is due to Dillon and Roselle (1969). The variable $l(t) = s_{h+1}(t) - 1 \geq 0$ is the number of integers "$h+1$" in a random permutation π_{t-1} generated by the specification $[s_1, \cdots, s_h, s_{h+1}(t) - 1]$. If $l(t) = 0$, Result (ii) is obviously true. Now consider $l(t) \geq 1$. Since "$h+1$" is the largest integer in the permutation π_{t-1}, following immediately after the integer "$h+1$" must be either a fall or a level. Hence the total number of falls and levels involving the integer "$h+1$" equals $l(t) = s_{h+1}(t) - 1$. □

Theorem 6.6 *Let S be a set of integers having specification $[s] = [s_1, \cdots, s_n]$, where $s_i \geq 1$ and $\sum_{i=1}^{n} s_i = N$. Then the exact distribution of the number of rises in a random permutation $\pi \in \Pi_N(S)$ is given by*

$$P(R_N(\pi) = k) = \boldsymbol{\xi}_0 \left(\prod_{t=1}^{N} \boldsymbol{M}_t([s]) \right) \boldsymbol{e}_k', \qquad (6.24)$$

where $\boldsymbol{\xi}_0 = (1, 0, \cdots, 0)$, and the transition probability matrices $\boldsymbol{M}_t([s])$, $t = 1, \cdots, N$, are given by

$$\boldsymbol{M}_t([s]) = \left(\begin{array}{c|c} \boldsymbol{A}_t & \boldsymbol{B}_t \\ \hline \boldsymbol{O} & \boldsymbol{I} \end{array} \right), \qquad (6.25)$$

with

$$
A_t = \begin{array}{c} 1 \\ 2 \\ \vdots \\ x \\ \vdots \\ N_t^* - 1 \end{array}
\left(
\begin{array}{ccccccc}
\frac{1+l(t)}{t} & \frac{t-l(t)-1}{t} & & & & & \\
& \frac{2+l(t)}{t} & \frac{t-l(t)-2}{t} & & & 0 & \\
& & \ddots & \ddots & & & \\
& & & \frac{x+l(t)}{t} & \frac{t-l(t)-x}{t} & & \\
& 0 & & & \ddots & \ddots & \\
& & & & & & \frac{N_t^*-1+l(t)}{t}
\end{array}
\right),
$$

an $(N_t^* - 1) \times (N_t^* - 1)$ *matrix, and*

$$
B_t = \begin{array}{c} 1 \\ 2 \\ \vdots \\ \vdots \\ N_t^* - 1 \end{array}
\left(
\begin{array}{cccccc}
0 & & 0 & & 0 & \\
\vdots & & \ddots & \ddots & & 0 \\
& & & \ddots & \ddots & \\
0 & & & & & \\
\frac{t-l(t)-N_t^*+1}{t} & 0 & \cdots & & 0 &
\end{array}
\right),
$$

an $(N_t^* - 1) \times (N^* - N_t^* + 1)$ *matrix.*

Proof. If $Y_{t-1} = x$ and the $(l(t) + 1)$-th copy of the integer of "$h + 1$" is inserted into the t gaps of the permutation π_{t-1}, then, following from Lemma 6.2, we have that

(i) the number of rises remains x ($Y_t = x$) when the integer "$h + 1$" was inserted into either the x gaps of rises or the $l(t)$ gaps of falls or levels associated with integers "$h + 1$" in π_{t-1}, and

(ii) the number of rises becomes $x + 1$ ($Y_t = x + 1$) when the integer "$h + 1$" was inserted into the $(t - x - l(t))$ gaps of falls or levels *not* associated with the integers "$h + 1$" in π_{t-1}.

Hence the transition probabilities associated with Y_t, $t = 1, \cdots, N$, are defined as follows:

$$
P(Y_t = y | Y_{t-1} = x) = \begin{cases}
\frac{x+l(t)}{t} & \text{if } y = x, 1 \leq x \leq N_t^* - 1 \\
\frac{t-x-l(t)}{t} & \text{if } y = x + 1, 1 \leq x \leq N_t^* - 1 \\
1 & \text{if } y = x, x \geq N_t^* \\
0 & \text{otherwise.}
\end{cases} \tag{6.26}
$$

The above equation yields the transition probability matrices given in Eq. (6.25), where the state space is $\Omega = \{1, \cdots, N^*\}$. It follows from the definition of Y_t that $P(R_N(\pi) = k) = P(Y_N = k)$, for all $k \in \Omega$. Hence $R_N(\pi)$ is finite Markov chain imbeddable, and Eq. (6.24) follows immediately from Theorem 2.1. $\qquad\square$

Corollary 6.1 *For $k = 1, 2, \cdots, N^*$, the Simon Newcomb numbers equal*

$$A([s], k) = \frac{N!}{\prod_{i=1}^{n} s_i!} \boldsymbol{\xi}_0 \left(\prod_{t=1}^{N} \boldsymbol{M}_t([s]) \right) \boldsymbol{e}_k'. \qquad (6.27)$$

With $A([s], 1) \equiv 1$, the recursive equation

$$A([s], k) = (N - l(N) - k + 1)A([s'], k - 1) + (k + l(N))A([s'], k) \quad (6.28)$$

holds for $k = 2, \cdots, N^*$, where the $[s']$-specification is the $[s]$-specification with deletion of one of the s_n integers "n".

Proof. Since there are $N!/(\prod_{i=1}^{n} s_i!)$ random permutations in $\Pi_N(S)$ and they are equally likely, Eq. (6.27) follows from the fact that the Simon Newcomb numbers $A([s], k)$ equal the product of $N!/(\prod_{i=1}^{n} s_i!)$ and $P(R_N(\pi) = k)$.

By definition, the first gap is a rise, and hence if there is only one rise in the permutation π, then the permutation has to have the form $\pi = (n \cdots n(n - 1) \cdots (n - 1) \cdots 1 \cdots 1)$ and $A([s], 1) \equiv 1$. The recursive Eq. (6.28) follows directly from Eqs. (6.24) and (6.25) of Theorem 6.6, since, for $k = 2, \cdots, N^*$,

$$\begin{aligned} P(R_N(\pi) = k) &= \boldsymbol{\xi}_0 \left(\prod_{t=1}^{N-1} \boldsymbol{M}_t([s]) \right) \left(\boldsymbol{M}_N([s]) \boldsymbol{e}_k' \right) \qquad (6.29) \\ &= \frac{N - l(N) - k + 1}{N} P(R_{N-1}(\pi) = k - 1) \\ &\quad + \frac{k + l(N)}{N} P(R_{N-1}(\pi) = k). \end{aligned}$$

$\qquad\square$

To make the insertion procedure and the above theorem more transparent, a simple example is given below.

Example 6.1 Let $S = \{1, 2, 2, 2, 3\}$. Then S has the specification $[s] = [1, 3, 1]$ with $N = 5$. The maximum number of rises in a random permutation $\pi \in \Pi_5(S)$ is $N^* = 3$. For $t = 2, 3, 4$, and 5, we have $N_2^* = 2$, $l(2) = 0$; $N_3^* = 2$, $l(3) = 1$; $N_4^* = 2$, $l(4) = 2$; and $N_5^* = 3$, $l(5) = 0$. It follows from Eq. (6.25) that the transition probability matrices associated with the finite Markov chain $\{Y_t\}$, $t = 2, \cdots, 5$, are

$$M_2([s]) = \begin{pmatrix} 1/2 & 1/2 & 0 \\ 0 & 1 & 0 \\ 0 & 0 & 1 \end{pmatrix}, \quad M_3([s]) = \begin{pmatrix} 2/3 & 1/3 & 0 \\ 0 & 1 & 0 \\ 0 & 0 & 1 \end{pmatrix},$$

$$M_4([s]) = \begin{pmatrix} 3/4 & 1/4 & 0 \\ 0 & 1 & 0 \\ 0 & 0 & 1 \end{pmatrix}, \quad M_5([s]) = \begin{pmatrix} 1/5 & 4/5 & 0 \\ 0 & 2/5 & 3/5 \\ 0 & 0 & 1 \end{pmatrix}.$$

With $M_1([s]) = I_{3\times 3}$ and $\boldsymbol{\xi}_0 = (1, 0, 0)$, therefore, for $1 \le k \le N^* = 3$,

$$P(R_5(\pi) = k) = \boldsymbol{\xi}_0 (\prod_{t=1}^{5} M_t) e_k' = (\frac{1}{20}, \frac{10}{20}, \frac{9}{20}) e_k'.$$

Hence $P(R_5(\pi) = 1) = 1/20$, $P(R_5(\pi) = 2) = 10/20$, and $P(R_5(\pi) = 3) = 9/20$. Since there are $20(=5!/3!)$ possible permutations in $\Pi_5(S)$, the Simon Newcomb numbers are $A([s], 1) = 1$, $A([s], 2) = 10$, and $A([s], 3) = 9$. This result can be checked by writing out the 20 permutations in $\Pi_5(S)$ and classifying them into three subsets of permutations, according to the number of rises (1, 2, or 3): (i) $\{(32221)\}$, (ii) $\{(31222), (32122), (32212), (21322), (22132), (22213), (23221), (22321), (22231), (13222)\}$, and (iii) $\{(12223), (12232), (12322), (21223), (21232), (22123), (23122), (23212), (22312)\}$. $\diamond$

For the case $s_1 = \cdots = s_n \equiv 1$, it is easy to check that (i) $N_t^* \equiv t$ and $l(t) \equiv 0$, for all $t = 1, 2, \cdots, n$, (ii) $\Omega = \{1, 2, \cdots, n\}$, and (iii) Eq. (6.26) is reduced to

$$P(Y_t = y | Y_{t-1} = x) = \begin{cases} x/t & \text{if } y = x, 1 \le x \le t-1 \\ (t-x)/t & \text{if } y = x+1, 1 \le x \le t-1 \\ 1 & \text{if } y = x, t \le x \le n \\ 0 & \text{otherwise.} \end{cases} \tag{6.30}$$

The following corollary gives the exact distribution of $R_n(\pi)$, and hence the Eulerian numbers.

Corollary 6.2   *Given the random permutation $\pi \in \Pi(Z_n)$:*

(i) The distribution of the number of rises $R_n(\pi)$ is given by

$$P(R_n(\pi) = k) = \boldsymbol{\xi}_0 (\prod_{t=1}^{n} M_t) e_k', \quad k = 1, 2, \cdots, n, \tag{6.31}$$

where $\boldsymbol{\xi}_0 = (1, 0, \cdots, 0)$, and the transition probability matrices $\boldsymbol{M}_t$, $t = 1, 2, \cdots, n$, equal

$$
\boldsymbol{M}_t =
\begin{array}{c}
1 \\ \vdots \\ k \\ \vdots \\ t-1 \\ t \\ \vdots \\ n
\end{array}
\left(
\begin{array}{ccccc|cc}
\frac{1}{t} & \frac{t-1}{t} & & & & & \\
& \ddots & \ddots & & & & \boldsymbol{O} \\
& & \frac{k}{t} & \frac{t-k}{t} & & & \\
& & & \ddots & \ddots & & \\
& & & & \frac{t-1}{t} & \frac{1}{t} & \\
\hline
& & \boldsymbol{O} & & & & \boldsymbol{I}
\end{array}
\right) . \qquad (6.32)
$$

(ii) The Eulerian number $A(n, k)$, the number of random permutations in $\Pi(Z_n)$ having exactly k rises, is given by

$$
A(n, k) = n! \boldsymbol{\xi}_0 \left(\prod_{t=1}^{n} \boldsymbol{M}_t \right) \boldsymbol{e}'_k, \qquad (6.33)
$$

and satisfies the recursive equation

$$
A(n, k) = k A(n-1, k) + (n - k + 1) A(n-1, k-1). \qquad (6.34)
$$

Proof. Since Z_n has specification $[s] = [1, 1, \cdots, 1]$, the first part of the theorem follows immediately from Theorem 6.6 and Eq. (6.30). The size of $\Pi(Z_n)$ is $n!$, and the result (6.33) follows directly from the fact that all permutations are equally likely with respect to the random insertion procedure.

Note that, by (i), the probability $P(R_n(\pi) = k)$ satisfies the recursive equation

$$
P(R_n(\pi) = k) =
\begin{cases}
\frac{1}{n} P(R_{n-1}(\pi) = 1) & \text{if } k = 1 \\
\frac{k}{n} P(R_{n-1}(\pi) = k) & \\
\quad + \frac{n-k+1}{n} P(R_{n-1}(\pi) = k-1) & \text{if } k = 2, \cdots, n-1 \\
\frac{1}{n} P(R_{n-1}(\pi) = n-1) & \text{if } k = n.
\end{cases}
$$
$$(6.35)$$

The recursive equation (6.34) is a direct consequence of $A(n, k) = n! P(R_n(\pi) = k)$ and Eq. (6.35). $\qquad \square$

The well-known Worpitzky (1883) identity for the Eulerian numbers, as given by Eq. (6.22), follows immediately from Eq. (6.33) and the identity

$$k\binom{x+k-1}{n} + (n-k)\binom{x+k}{n} = x\binom{x+k-1}{n-1}, \quad k=1,\cdots,n,$$

as shown below:

$$
\begin{aligned}
\sum_{k=1}^{n}\binom{x+k-1}{n}A(n,k) &= (n-1)!\boldsymbol{\xi}_0\prod_{t=1}^{n-1}\boldsymbol{M}_t\begin{pmatrix}\binom{x}{n}+(n-1)\binom{x+1}{n}\\2\binom{x+1}{n}+(n-2)\binom{x+2}{n}\\\vdots\\(n-1)\binom{x+n-2}{n}+\binom{x+n-1}{n}\\n\binom{x+n-1}{n}\end{pmatrix}\\[2mm]
&= (n-1)!\boldsymbol{\xi}_0\prod_{t=1}^{n-1}\boldsymbol{M}_t\begin{pmatrix}\binom{x}{n-1}\\\binom{x+1}{n-1}\\\vdots\\\binom{x+n-2}{n-1}\\\binom{x+n-1}{n-1}\end{pmatrix}\cdot x\\[2mm]
&= x^{n-1}\cdot x = x^n.
\end{aligned}
\tag{6.36}
$$

Further, Euler's original form of the definition of $A(n,k)$ given in Eq. (6.19) can also be derived from the method of finite Markov chain imbedding. Note that the Eulerian polynomial satisfies the equation

$$
\begin{aligned}
A_n(x) &= \sum_{k=1}^{n}A(n,k)x^k\\[2mm]
&= n!\boldsymbol{\xi}_0\prod_{t=1}^{n-1}\boldsymbol{M}_t\begin{pmatrix}\frac{1}{n}&\frac{n-1}{n}&&&&\\&\frac{2}{n}&\frac{n-2}{n}&&\boldsymbol{O}&\\&&\ddots&\ddots&&\\&&&\frac{k}{n}&\frac{n-k}{n}&\\&\boldsymbol{O}&&&\ddots&\\&&&&&1\end{pmatrix}\begin{pmatrix}x^1\\x^2\\\vdots\\x^k\\\vdots\\x^n\end{pmatrix}\\[2mm]
&= nxA_{n-1}(x) + x(1-x)DA_{n-1}(x),
\end{aligned}
\tag{6.37}
$$

where $D = d/dx$, $A_0(x) \equiv 1$ for all x, and its exponential generating function,

$$\varphi(x,\lambda) = \sum_{n=0}^{\infty} A_n(x)\frac{\lambda^n}{n!}, \tag{6.38}$$

satisfies the differential equation

$$(1 - x\lambda)\varphi_\lambda(x,\lambda) = x\varphi(x,\lambda) + x(1-x)\varphi_x(x,\lambda),$$

where $\varphi_x(x,\lambda) = \partial\varphi(x,\lambda)/\partial x$ and $\varphi_\lambda(x,\lambda) = \partial\varphi(x,\lambda)/\partial\lambda$ with boundary conditions $\varphi(x,0) = \varphi(0,\lambda) \equiv 1$. Solving the above linear partial differential equation by Lagrange's method yields the exponential generating function

$$\varphi(x,\lambda) = \frac{1-x}{1 - x\exp\{\lambda(1-x)\}}. \tag{6.39}$$

Identity (6.19) follows naturally from Eqs. (6.37)-(6.39).

Similarly, we could find the distribution for the number of falls (decreases), $D_N(\pi)$, the same way in which we found the distribution for $R_N(\pi)$. For the special case when $[s] = [1,1,\cdots,1]$, we have $R_n(\pi) + D_n(\pi) \equiv n+1$. This yields that the distribution of $D_n(\pi)$ is given by

$$P(D_n(\pi) = k) = P(R_n(\pi) = n-k+1), \quad k = 1,\cdots,n,$$

and that their joint distribution is given by

$$P(D_n(\pi) = k, R_n(\pi) = n-k+1) = P(R_n(\pi) = n-k+1),$$

for $k = 1,\cdots,n$, and 0 otherwise.

For general cases, with some $s_i \geq 2$, the joint distribution of $R_N(\pi)$ and $D_N(\pi)$ is rather complex (see Fu and Lou 2000b).

Chapter 7

Applications

7.1 Introduction

The application of the finite Markov chain imbedding technique started in the early 1980s for the purpose of evaluating the reliabilities of various engineering systems, especially the consecutive-k-out-of-n:F system. Since then, it has been used, for example, in hypothesis testing by Lou (1996, 1997), Koutras and Alexandrou (1997a), and Johnson (2001), in quality control by Chao (1999) and Fu, Spiring and Xie (2002), in continuity of care by Lou (2000, 2001) and Fu and Lou (2000a), and in DNA sequence matching by Fu, Lou and Chen (1999), Cheung (2002), and Lou (2003). In this chapter, our main goal is to present some of these works in detail, and to provide a scope for further applications of this approach to other areas.

7.2 System Reliability

Engineering systems such as atomic power plants, aircrafts, automobiles, computers, and software programs are required to be highly reliable. Reliability evaluation has become an important and integral part of the planning, design and operation of all types of engineering systems. There are various reliability structures, including series systems, parallel systems, k-out-of-n:F systems, consecutive-k-out-of-n:F systems, linear systems, and deteriorating and repairable systems. It is widely known that the reliability of series systems is low especially when the number of components in the system is large, and, on the other hand, parallel systems have high reliability but are expensive to build. Recently, consecutive-k-out-of-n:F systems have become very popular for their high reliability and relatively low cost; the analysis of these systems will be the focus of this section.

Let $\{X_i\}_{i=1}^n$ be a sequence of two-state indicator random variables for the proper functioning of n components: $X_i = 1$ if and only if the i-th component works and $X_i = 0$ otherwise. Let $\phi(X_1, \cdots, X_n)$ be a two-state function defined on $\{X_i\}$ with $\phi(X_1, \cdots, X_n) = 1$ if and only if the system works and zero otherwise. The indicator function $\phi(X_1, \cdots, X_n)$ is commonly referred to as the structure function of the system. For example, a series system can be represented by $\phi(X_1, \cdots, X_n) = \min\{X_1, \cdots, X_n\}$, while $\phi^*(X_1, \cdots, X_n) = \max\{X_1, \cdots, X_n\}$ denotes a parallel system. It is well known that for any system, the structure function ϕ can be written in the form

$$\phi(X_1, \cdots, X_n) = \min_j \{\phi_j(X_1, \cdots, X_n)\}, \tag{7.1}$$

where ϕ_j is the structure function of a parallel system (see, for example, Ross 2000). This means that any complex system can be decomposed into a series system with each component equivalent to a parallel subsystem. The reliability of the system, R_s, is the probability that the system is functioning, or, equivalently,

$$R_s = E(\phi(X_1, \cdots, X_n)). \tag{7.2}$$

Although the structure function in Eq. (7.1) and the reliability function in Eq. (7.2) are presented in simple mathematical form, they are unworkable for most cases, especially when the structure functions ϕ_j are complex, as in, for example, consecutive-k-out-of-n:F systems (denoted $C(k, n\!:\!F)$), deteriorating and repairable systems, or if the components X_i are multi-state Markov-dependent trials.

The reliability of the $C(k, n\!:\!F)$ system was studied extensively during the 1980s via combinatorial analysis by, for example, Kontoleon (1980), Chiang and Niu (1981), Hwang (1982, 1986), Papastavridis (1988), Chrysaphinou and Papastavridis (1988), and Kossow and Preuss (1989). From the point of view of the theory of runs and patterns, the $C(k, n\!:\!F)$ system is a linear system containing n components, and the system works if and only if no pattern (or failure run) $\Lambda = F \cdots F$ of size k has occurred in the sequence of n two-state trials. Mathematically, the reliability of the $C(k, n\!:\!F)$ system is

$$R(k, n\!:\!q) = P(W(\Lambda) \geq n + 1). \tag{7.3}$$

More generally, the reliability of any engineering system with n components equals the tail probability of a waiting-time random variable $W(\Lambda)$ of a specific pattern Λ (simple, compound, or series). This fundamental connection paves a simple and effective way for evaluating the reliability using the finite Markov chain imbedding technique, without the usual strong restrictions that the components have to be independent and identically distributed.

7.2.1 *Consecutive-k-out-of-n:F System*

The following two examples provide descriptions of $C(k, n\!:\!F)$ systems (Chiang and Niu 1981, and Chao, Fu and Koutras 1995).

Example 7.1 A sequence of n microwave stations transmit information from place A to B. The microwave stations are equally spaced between A and B. Each microwave station is able to transmit information a distance up to k microwave stations away. This system fails if and only if k or more consecutive microwave stations fail. $\diamond$

Example 7.2 A system for transporting oil by pipes from point A to B has n pump stations. The pump stations are equally spaced between A and B, and each pump station can transport the oil a distance of k pump stations away. If one pump station is down, the flow of oil would not be interrupted because the next station could carry the load. However, when k or more consecutive pump stations fail, the oil flow stops and the system fails. $\diamond$

Kontoleon (1980) studied the reliability of the $C(k, n\!:\!F)$ system where all the components are stochastically independent and have the same failure probability (*i.e.* Bernoulli trials). For such a system, Chiang and Niu (1981) provide the recursive equation

$$R(k, n\!:\!q) = p^{n-k+1} + \sum_{r=1}^{n-k+1} \sum_{m=r+1}^{r+k-1} R(k, n-m:q)p^r q^{m-r}, \qquad (7.4)$$

with $R(k, j\!:\!q) = 1$ if $0 \le j < k$. Derman, Lieberman, and Ross (1982) expressed $R(k, n\!:\!q)$ in the form

$$R(k, n\!:\!q) = \sum_{j=0}^{n} N(n, j, k)p^{n-j}q^j, \qquad (7.5)$$

where the coefficients $N(n, j, k)$ are referred to as Fibonacci numbers of order k. For *i.i.d.* cases, Philippou and Makri (1986) and Hirano (1986) independently developed the exact formula for the distribution of the random variable $N_{n,k}$ (see Eq. (3.1) of Section 3.2), and the reliability of a $C(k, n\!:\!F)$ system equals the probability that $N_{n,k} = 0$.

In a series of papers by Fu (1985, 1986), Fu and Hu (1987), and Chao and Fu (1989, 1991), the finite Markov chain imbedding technique, with forward and backward principle, was introduced to study the reliability of the $C(k, n\!:\!F)$ system, its bounds and its limiting distributions. We summarize below the main results of these papers.

Theorem 7.1   *Assuming all the components operate independently and have the same failure probability q, then*

(i)

$$R(k, n\!:\!q) = \boldsymbol{\xi}\boldsymbol{N}^n(\Lambda)\mathbf{1}', \tag{7.6}$$

where the $\Lambda = F \cdots F$ is a simple pattern of length k and the matrix $\boldsymbol{N}(\Lambda)$ has the form

$$\boldsymbol{N}(\Lambda) = \begin{matrix} 0 \\ 1 \\ 2 \\ \vdots \\ k-1 \end{matrix} \begin{pmatrix} p & q & & & \\ p & 0 & q & & 0 \\ p & 0 & 0 & \ddots & \\ \vdots & \ddots & \ddots & \ddots & q \\ p & & 0 & \cdots & 0 \end{pmatrix}_{k \times k}, \tag{7.7}$$

(ii)

$$(1 - q^k)^{n-k+1} \le R(k, n\!:\!q) \le (1 - pq^k)^{n-k+1}, \tag{7.8}$$

and

(iii)

$$R(k, n\!:\!q) \sim \exp\{n \log \lambda_{[1]}\}, \tag{7.9}$$

where $\lambda_{[1]}$ $(0 < \lambda_{[1]} < 1)$ is the largest eigenvalue of $\boldsymbol{N}(\Lambda)$.

With some simple modifications to the transition probability matrices, the results (i) and (iii) of Theorem 7.1 also hold for Markov-dependent sequences of components (see Chapter 3 for simple patterns). Inequality (7.8) has been studied by Fu (1985, 1986), Cai (1994), Papastavridis and

Koutras (1993), and recently by Muselli (2000). We would like to point out an important connection between the lower bound of the inequality (7.8) and the Poisson convergence, when q is very small.

Assuming the failure probability of each component within the period $[0, s]$ is $q = (\lambda s/n)^{1/k}$, then it follows from the inequality (7.8) that

$$\lim_{n \to \infty} R(k, n : q) = \exp\{-\lambda s\}, \quad \lambda, s > 0. \tag{7.10}$$

This implies that, if q is sufficiently small for a given large n, the reliability of the system in the time period $[0, s]$ is approximately

$$R(k, n : q) \sim \exp\{-nq^k\}. \tag{7.11}$$

In other words, if once the system breaks down it will be replaced immediately by a new system, then the number of times of failure of the system $N_s(\Lambda)$ during the period $[0, s]$ has a Poisson distribution in the sense that

$$P(N_s(\Lambda) \leq x) \sim \sum_{j=0}^{x} \frac{(\lambda s)^j}{j!} \exp\{-\lambda s\}, \tag{7.12}$$

as $q = (\lambda s/n)^{1/k} \to 0$. Results (7.10) and (7.12) have been studied by, for example, Chao and Lin (1984), Fu (1986), Papastavridis (1988), Godbole (1991), and Koutras and Papastavridis (1993).

For fixed failure probability q, we would also like to point out an interesting connection between the upper bound $(1 - pq^k)^{n-k+1}$ of the inequality (7.8) and the famous Goncharov (1944) approximation for the longest failure run: for large n

$$R(k, n : q) = P(L_n(F) < k) \sim \exp\{-\lambda_n[1 + o(1)]\}, \tag{7.13}$$

where $\lambda_n = npq^{[\log_{1/q} n] + x}$ and $x = k - [\log_{1/q} n]$. To see this connection, let us take $k = [\log_{1/q} n] + x$, then the upper bound of the inequality (7.8) yields

$$
\begin{aligned}
(1 - pq^k)^{n-k+1} &= (1 - pq^{[\log_{1/q} n] + x})^{n - [\log_{1/q} n] - x + 1} \\
&= (1 - \lambda_n/n)^{n(1 + o(1))} \\
&\cong \exp\{-\lambda_n(1 + o(1))\}.
\end{aligned}
$$

It is well known that $L_n(F) - [\log_{1/q} n]$ does not have a limiting distribution (see Goncharov 1944 and Vaggelatou 2003), meaning that Eq. (7.13) is only a local approximation along the sequence $\{[\log_{1/q} n] + x, x = 0, \pm 1, \pm 2, \cdots\}$.

Hence we don't expect this approximation to perform well for fixed k and large n. Further, since the Poisson approximation (7.11) for $R(k, n : q)$ requires $q = (\lambda s/n)^{1/k}$ tending to zero as $n \to \infty$, we also don't expect it to perform well numerically for fixed q and large n. If q is very small (and hence p is close to 1), then the lower and upper bounds of the inequality (7.8) are almost equal, hence providing a good approximation for the exact reliability $R(k, n:q)$. In other words, for small q, the lower bound of Eq. (7.8) performs better than the Poisson approximation given by Eq. (7.11).

We would first like to provide some numerical results for Theorem 7.1 before providing its proof. Let us introduce the following notation:

E: The exact reliability $R(k, n:q)$ in Eq. (7.6),
U: The upper bound of $R(k, n:q)$ in Eq. (7.8),
L: The lower bound of $R(k, n:q)$ in Eq. (7.8),
A_L: The large deviation approximation of $R(k, n:q)$ in Eq. (7.9),
A_G: The Goncharov approximation of $R(k, n:q)$ in Eq. (7.13).

Table 7.1 gives the exact probability $R(k, n:q)$ and its various bounds and approximations for *i.i.d.* two-state trials with $q = 0.1$ and $k = 4$, and Table 7.2 shows the effect of increasing the failure probability to $q = 0.3$.

Table 7.1   Comparisons of the exact reliability $R(k, n:q)$ to bounds and approximations for *i.i.d.* two-state trials with $q = 0.1$ and $k = 4$.

n	L	A_G	E	A_L	U
50	0.9950	0.9955	0.9958	0.9955	0.9958
100	0.9901	0.9910	0.9913	0.9910	0.9913
1000	0.9048	0.9139	0.9141	0.9139	0.9142
10000	0.3679	0.4066	0.4065	0.4064	0.4067

In Table 7.3, we provide some numerical results for the reliability of the $C(k, n:F)$ system under a Markov-dependent structure of components having the transition probability matrix

$$A^* = \begin{pmatrix} 0.75 & 0.25 \\ 0.25 & 0.75 \end{pmatrix}.$$

Summarizing the numerical results, for q small $(q = 0.1)$, all the bounds and approximations perform reasonably well, and for a more moderate value

Table 7.2 Comparisons of the exact reliability $R(k, n:q)$ to bounds and approximations for *i.i.d.* two-state trials with $q = 0.3$ and $k = 4$.

n	L	A_G	E	A_L	U
50	0.6659	0.7531	0.7590	0.7475	0.7654
100	0.4434	0.5672	0.5674	0.5588	0.5761
1000	0.0003	0.0035	0.0030	0.0030	0.0035
10000	4.77e-36	2.37e-25	5.35e-26	5.27e-26	2.06e-25

Table 7.3 Comparisons of the exact reliability $R(k, n:q)$ to the large deviation approximation for Markov-dependent two-state trials with transition probability matrix A^* and initial probabilities $p_0 = q_0 = 0.5$.

n	$k = 4$		$k = 8$	
	E	A_L	E	A_L
20	0.2094	0.1841	0.7375	0.6673
50	0.0165	0.0145	0.4020	0.3637
100	2.40e-04	2.11e-04	0.1462	0.1323
1000	2.01e-37	1.77e-37	1.81e-09	1.64e-09
10000			1.55e-88	1.40e-88

of q ($q = 0.3$), the upper bound U and the approximations A_G and A_L still perform very well for small and moderate n. Not surprisingly, for large n, A_L outperforms other approximations and bounds, regardless of the values of q and k; as the number of components n increases, the reliability of the system $R(k, n:q)$ tends to zero exponentially at the same rate as A_L (see Theorems 5.9 and 5.10). Since the computation of $R(k, n:q)$ based on Eq. (7.6) is very simple and also very efficient, we believe that the reliability $R(k, n:q)$ should generally be evaluated by the exact formula in Eq. (7.6), and the large deviation approximation should be used only if the number of components n is very large.

The result of Theorem 7.1(i) follows immediately from the fact that the $C(k, n:F)$ system works if and only if $N_{n,k} = 0$ or $L_n(F) < k$. Theorem 7.1(iii) can be proved in the exact same manner as Theorem 5.9. Hence to prove Theorem 7.1 we need only prove Part (ii), the upper and lower bounds of the system reliability.

Let $\boldsymbol{a}(0) = \boldsymbol{\xi} = (1, 0, \cdots, 0)_{1 \times k}$, and for $i = 1, 2, \cdots, n$, let

$$\boldsymbol{a}(i) = \boldsymbol{\xi} \boldsymbol{N}^i = (a_0(i), a_1(i), \cdots, a_{k-1}(i)), \qquad (7.14)$$

where $a_j(i)$ is the probability of the system residing in state j at time index $t = i$ (after the first i components are counted). In component form, we write

$$a_j(i) = \boldsymbol{\xi} \boldsymbol{N}^i \boldsymbol{e}'_{j+1}, \quad j = 0, 1, \cdots, k - 1. \qquad (7.15)$$

Note that in the above Eq. (7.15), we have $\boldsymbol{e}'_{j+1}$ rather than $\boldsymbol{e}'_j$. This is because while j is the number of failures and starts from 0, here we wish to emphasize that j corresponds to the $(j + 1)$-th state in the state space Ω. It follows from the structure of the transition probability matrix in Eq. (7.7) and from Eq. (7.15) that the following recursive equations hold: for $1 \le i \le n$,

$$\begin{aligned} a_0(i) &= p[a_0(i-1) + a_1(i-1) + \cdots a_{k-1}(i-1)] \text{ and} \\ a_j(i) &= q a_{j-1}(i-1), \quad j = 1, \cdots, k - 1. \end{aligned} \qquad (7.16)$$

To prove Theorem 7.1(ii), we introduce the following three lemmas.

Lemma 7.1

   (i) $\boldsymbol{N}\boldsymbol{1}' = \boldsymbol{1}' - q\boldsymbol{e}'_k$,
   (ii) $\boldsymbol{N}\boldsymbol{e}'_1 = p\boldsymbol{1}'$,
   (iii) $\boldsymbol{N}\boldsymbol{e}'_l = q\boldsymbol{e}'_{l-1}, \; l = 2, \cdots, k$.

Proof. The above results follow directly from the structure of the matrix $\boldsymbol{N}$ given by Eq. (7.7). $\qquad\qquad\qquad\qquad\qquad\qquad\qquad\qquad\qquad$ □

Lemma 7.2

   (i) $\boldsymbol{a}(i)\boldsymbol{1}' = \boldsymbol{a}(i-1)\boldsymbol{1}' - q\boldsymbol{a}(i-1)\boldsymbol{e}'_k$,
   (ii) $\boldsymbol{a}(i)\boldsymbol{1}' \le \boldsymbol{a}(j)\boldsymbol{1}'$, *for all* $0 \le j \le i \le n$,
   (iii) $\boldsymbol{a}(i-k+1)\boldsymbol{e}'_1 \le \boldsymbol{a}(i)\boldsymbol{1}'$, *for all* $i \ge k$.

Proof. Result (i) of this lemma is a direct consequence of Lemma 7.1(i) and the definition of $\boldsymbol{a}(i)$ given by Eq. (7.14). Since $q\boldsymbol{a}(i-1)\boldsymbol{e}'_k \ge 0$, Result (ii) of this lemma follows directly from (i). Finally, Result (iii) follows from

$$\begin{aligned} \boldsymbol{a}(i)\boldsymbol{1}' &= \boldsymbol{a}(i-1)(\boldsymbol{1}' - \boldsymbol{e}'_k) + p\boldsymbol{a}(i-1)\boldsymbol{e}'_k \\ &\ge \boldsymbol{a}(i-1)(\boldsymbol{1}' - \boldsymbol{e}'_k) \end{aligned}$$

$$= \mathbf{a}(i-2)(\mathbf{1}' - \mathbf{e}'_k - \mathbf{e}'_{k-1}) + p\mathbf{a}(i-2)\mathbf{e}'_{k-1}$$
$$\geq \mathbf{a}(i-2)(\mathbf{1}' - \mathbf{e}'_k - \mathbf{e}'_{k-1})$$
$$\vdots$$
$$= \mathbf{a}(i-k+1)(\mathbf{1}' - \mathbf{e}'_k - \cdots - \mathbf{e}'_2) + p\mathbf{a}(i-k+1)\mathbf{e}'_2$$
$$\geq \mathbf{a}(i-k+1)\mathbf{e}'_1.$$

□

The above inequalities (ii) and (iii) are critical to proving the inequality (7.8). We would like to give the intuitive implications of these two inequalities:

(A) The inequality of Lemma 7.2(ii) implies that $R(k, n:q)$ is a decreasing function of the number of components.
(B) The inequality of Lemma 7.2(iii) shows that the reliability of the $C(k, n:F)$ system with the first i ($i \geq k$) components is greater than or equal to the probability that the system with the first $(i-k+1)$ components is at the state "0" (no failures).

Lemma 7.3 *For $n \geq i \geq k$, we have*

$$pq^{k-1}\mathbf{a}(i)\mathbf{1}' \leq \mathbf{a}(i)\mathbf{e}'_k \leq q^{k-1}\mathbf{a}(i)\mathbf{1}'. \tag{7.17}$$

Proof. Since

$$\mathbf{a}(i)\mathbf{e}'_k = q^{k-1}\mathbf{a}(i-k+1)\mathbf{e}'_1 = pq^{k-1}\mathbf{a}(i-k)\mathbf{1}',$$

Eq. (7.17) follows immediately from the inequalities of Lemma 7.2(ii) and (iii).

□

Proof. [of Theorem 7.1(ii)]. By Lemma 7.2,

$$\begin{aligned} R(k, n:q) &= \boldsymbol{\xi} \mathbf{N}^n \mathbf{1}' = \mathbf{a}(n)\mathbf{1}' = \mathbf{a}(n-1)\mathbf{1}' - q\mathbf{a}(n-1)\mathbf{e}'_k \\ &= \left(1 - \frac{q\mathbf{a}(n-1)\mathbf{e}'_k}{\mathbf{a}(n-1)\mathbf{1}'}\right)\mathbf{a}(n-1)\mathbf{1}' \\ &\vdots \\ &= \prod_{i=1}^{n}\left(1 - \frac{q\mathbf{a}(n-i)\mathbf{e}'_k}{\mathbf{a}(n-i)\mathbf{1}'}\right)\mathbf{a}(0)\mathbf{1}'. \end{aligned}$$

Note that $\boldsymbol{a}(0) = \boldsymbol{\xi} = (1, 0, \cdots, 0)$ and $\boldsymbol{a}(0)\mathbf{1}' \equiv 1$. Hence

$$R(k, n : q) = \prod_{i=1}^{n} \left(1 - \frac{q\boldsymbol{a}(n-i)\boldsymbol{e}_k'}{\boldsymbol{a}(n-i)\mathbf{1}'} \right). \tag{7.18}$$

Applying Lemma 7.3 to Eq. (7.18) yields the inequality

$$(1 - q^k)^{n-k+1} \leq R(k, n : q) \leq (1 - pq^k)^{n-k+1}.$$

$\square$

7.2.2 *Linearly Connected System*

An engineering system having n components is referred to as a *linearly connected system* if it can be imbedded into a finite Markov chain $\{Y_t : t = 0, 1, \cdots, n\}$ defined on a finite state space $\Omega = \{0, 1, \cdots, k-1, \alpha\}$ with transition probability matrices

$$\boldsymbol{M}(t; n) = \begin{matrix} 0 \\ \vdots \\ k-1 \\ \alpha \end{matrix} \begin{pmatrix} p_{0,0}(t; n) & \cdots & p_{0,\alpha}(t; n) \\ \vdots & \cdots & \vdots \\ p_{k-1,0}(t; n) & \cdots & p_{k-1,\alpha}(t; n) \\ 0 & \cdots & 1 \end{pmatrix} = \left(\begin{array}{c|c} \boldsymbol{N}(t; n) & \boldsymbol{C}(t; n) \\ \hline \mathbf{0} & 1 \end{array} \right).$$

$$\tag{7.19}$$

State α is an absorbing state at which the system breaks down and cannot be used anymore.

Let $\boldsymbol{\xi}_0 = (\boldsymbol{\xi} : 0)$ be the initial distribution of Y_0. Since $\{Y_t\}$ is a Markov chain and α is an absorbing state, it follows that the reliability of the linearly connected system can be represented as

$$
\begin{aligned}
R_l(k, n : q) &= P(Y_0 \leq k-1, Y_1 \leq k-1, \cdots, Y_n \leq k-1) \\
&= P(Y_0 \leq k-1)P(Y_1 \leq k-1 | Y_0 \leq k-1) \cdots \\
&\quad \cdots P(Y_n \leq k-1 | Y_{n-1} \leq k-1) \\
&= \boldsymbol{\xi} \left(\prod_{t=1}^{n} \boldsymbol{N}(t; n) \right) \mathbf{1}'.
\end{aligned}
\tag{7.20}
$$

The structure of the linearly connected system is very general; it covers many well known systems such as series, standby, k-out-of-n:F, consecutive-k-out-of-n:F, m-within-consecutive-k-out-of-n:F, and deteriorating and repairable systems. The paper by Chao, Fu and Koutras (1995) provides a more detailed review of the reliabilities for such systems.

7.3 Hypothesis Testing

The formal application of the distribution theory of runs and patterns to hypothesis testing started with Wald and Wolfowitz (1940). They proposed a test based on the conditional distribution of the total number of runs given the number of successes in Bernoulli trials. The null distribution of the test was studied by Swed and Eisenhart (1943), and the power functions were studied by David (1947), Bateman (1948), Barton and David (1958), and Goodman (1958). Since then, many tests based on runs and patterns have been proposed; for example, Rubin, McCulloch, and Shapiro (1990) proposed a test, based on the number of runs in multinomial data, of randomness versus clustering. The main purpose of this section is to show how the finite Markov chain imbedding technique could be used in computing critical regions and powers in a runs-related test.

One of the most commonly used runs-related statistics for testing randomness is the conditional test based on the number of success runs given the number of successes $N_1 = n_1$ in a sequence of two-state trials. The unconditional distribution for the number of success runs R_n, and the distribution for the number of successes N_1, can be derived in a manner similar to what is described in the following, and hence we leave this to the reader. Here we focus on the conditional distribution of the number of success runs given the number of successes, $R_n | N_1 = n_1$, as it is commonly used in the literature.

The conditional distribution of R_n given $N_1 = n_1$ can be obtained directly through the joint distribution of R_n and N_1. This method can be applied not only to *i.i.d.* cases, but also to one-step Markov-dependent cases, and hence can be used to calculate both the critical regions and powers of the success runs test (Lou 1996, 1997).

For a sequence of two-state trials $\omega = (z_1, \cdots, z_n)$ with n_1 successes, define

    (a) a Markov chain $\{Y_t\}$ operating on ω: $Y_t(\omega) = (B, R, E)$, where B is the number of successes in the first t trials, R is the number of success runs in the first t trials, and E is the outcome of the t-th trial (used to help identify the transition probabilities),

    (b) the state space $\Omega = (0, 0, F) \cup \{(b, r, e) : b = 1, \cdots, n_1; \ r = 1, \cdots, m; e = S, F\} \cup \alpha$, where $m = \min(n_1, n_2 + 1)$, $r \leq b$, α is an absorbing state (standing for states with $b > m$ or $r > m$), and

$k \equiv Card(\Omega) = 2 + m(m+1)$ if $n_1 = m$ or $k = 2 + m(2n_1 - m + 1)$
if $n_1 > m$,

(c) the partitions $C_0 = \{(0,0,F)\}$, $C_\alpha = \{\alpha\}$, $C_{n_1,r} = \{(n_1,r,S),(n_1,r,F)\}$, and $C_{b,r} = \{(b,r,S),(b,r,F)\}$, where $b = 1,\cdots, n_1 - 1$, and $r = 1,\cdots,m$, with $r \le b$.

Theorem 7.2 *Consider a sequence of n homogeneous Markov-dependent two-state trials. For given n and n_1, the random vector (N_1, R_n) can be imbedded into the Markov chain $\{Y_t\}$ defined above with transition probability matrix*

$$
M = \begin{array}{c}
(0,0,F) \\ (1,1,S) \\ (1,1,F) \\ (2,1,S) \\ \vdots \\ (n_1,m,F) \\ \alpha
\end{array}
\begin{pmatrix}
p_{FF} & p_{FS} & 0 & 0 & \cdots & 0 \\
0 & 0 & p_{SF} & p_{SS} & \cdots & 0 \\
0 & 0 & p_{FF} & 0 & \cdots & \\
\vdots & \vdots & \ddots & \ddots & \ddots & \\
 & \vdots & & \ddots & \ddots & \ddots \\
 & & & \ddots & \ddots & \cdot \\
0 & \cdots & 0 & & \cdots & 0 & 1
\end{pmatrix} (\star) \quad \forall t, \quad (7.21)
$$

where $(\star)$ stands for the probabilities $P(Y_t = \alpha|Y_{t-1} = i)$, $t = 1,\cdots,n$, $i \in \Omega - \{\alpha\}$. Given the initial condition $P(Y_0 = (0,0,F)) \equiv 1$, then the following results hold:

(1) For $n_1 = 0$ and 1, $P(N_1 = 0, R_n = 0) = P(N_1 = 1, R_n = 1) \equiv 1$, and for $n_1 \ge 2$, $r = 1,\cdots,m$,

$$P(N_1 = n_1, R_n = r) = \xi_0 M^n U^{'}(C_{n_1,r}), \quad (7.22)$$

where $\xi_0 = (1,0,\cdots,0)$ is a $1 \times k$ vector, k being the size of the state space.

(2) For $n_1 \ge 2$, $r = 1,\cdots,m$,

$$P(R_n = r|N_1 = n_1) = \frac{\xi_0 M^n U^{'}(C_{n_1,r})}{\xi_0 M^n U^{'}(C_{n_1})}, \quad (7.23)$$

where $C_{n_1} = \cup_{r=1}^m C_{n_1,r}$ and $U^{'}(C_{n_1})$ is the sum of all vectors $U^{'}(C_{n_1,r})$ over r.

Proof.   To prove the above results, it is sufficient to prove that the transition probability matrix M has the form given in Eq. (7.21).

For $n_1 = 0$ or 1, the result is obvious. For $n_1 \geq 2$, it follows from the constructions and definitions (a), (b), and (c), that the transition probabilities of the Markov chain $\{Y_t\}$, $t = 1, \cdots, n$, are given by

(i) for $1 \leq b \leq n_1$, $1 \leq r \leq m = \min(n_1, n_2 + 1)$,

$$P(Y_t = (b, r, F)|Y_{t-1} = (b, r, F)) = p_{FF},$$
$$P(Y_t = (b, r, F)|Y_{t-1} = (b, r, S)) = p_{SF},$$

(ii) for $1 \leq b \leq n_1 - 1$, $1 \leq r \leq m - 1$,

$$P(Y_t = (b + 1, r + 1, S)|Y_{t-1} = (b, r, F)) = p_{FS},$$
$$P(Y_t = (b + 1, r, S)|Y_{t-1} = (b, r, S)) = p_{SS},$$
$$P(Y_t = (b + 1, m, S)|Y_{t-1} = (b, m, S)) = p_{SS},$$

(iii) for $b = n_1$, $1 \leq r \leq m$,

$$P(Y_t = \alpha|Y_{t-1} = (b, r, F)) = p_{FS},$$
$$P(Y_t = \alpha|Y_{t-1} = (b, r, S)) = p_{SS},$$

(iv) $P(Y_t = \alpha|Y_{t-1} = \alpha) \equiv 1$, and zero otherwise.

Hence the transition probability matrix M is an immediate consequence of conditions (i) to (iv). Equations (7.22) and (7.23) follow directly from Theorem 2.1 and the definition of conditional probability. $\qquad\square$

To illustrate the theorem, we give the following example.

Example 7.3 For $n = 5$, $n_1 = 2$, $m = \min(2, 4) = 2$, the state space can be defined by $\Omega = \{(0, 0, F), (1, 1, S), (1, 1, F), (2, 1, S), (2, 1, F), (2, 2, S), (2, 2, F), \alpha\}$, where $(0, 0, F)$ is the initial state and stands for no success, and α represents states with more than two successes ($b > 2$). A Markov chain $\{Y_t\}$ is then defined. For instance, for a given sequence $\omega = (SFSFF)$, the realization of the Markov chain $Y_t(\omega)$, $t = 0, 1, \cdots, 5$, is $\{Y_0(\omega) = (0, 0, F), Y_1(\omega) = (1, 1, S), Y_2(\omega) = (1, 1, F), Y_3(\omega) = (2, 2, S), Y_4(\omega) = Y_5(\omega) = (2, 2, F)\}$.

If the two-state trials are homogeneous Markov-dependent, then the transition probability matrices of the imbedded Markov chain $\{Y_t\}$ are the same for all t, and can be expressed as

$$
M = \begin{array}{c} (0,0,F) \\ (1,1,S) \\ (1,1,F) \\ (2,1,S) \\ (2,1,F) \\ (2,2,S) \\ (2,2,F) \\ \alpha \end{array}
\left(
\begin{array}{cccccccc}
p_{FF} & p_{FS} & 0 & 0 & 0 & 0 & 0 & 0 \\
0 & 0 & p_{SF} & p_{SS} & 0 & 0 & 0 & 0 \\
0 & 0 & p_{FF} & 0 & 0 & p_{FS} & 0 & 0 \\
0 & 0 & 0 & 0 & p_{SF} & 0 & 0 & p_{SS} \\
0 & 0 & 0 & 0 & p_{FF} & 0 & 0 & p_{FS} \\
0 & 0 & 0 & 0 & 0 & 0 & p_{SF} & p_{SS} \\
0 & 0 & 0 & 0 & 0 & 0 & p_{FF} & p_{FS} \\
0 & 0 & 0 & 0 & 0 & 0 & 0 & 1
\end{array}
\right).
$$

The partition on the state space Ω for $n_1 = 2$ is generated by $C_{n_1,r} = \{(2,r,S),(2,r,F)\}$, $r = 1,2$. Then the exact probability of $R_{5,2}$ (R_{n,n_1} denotes the conditional random variable of R_n given $N_1 = n_1$) can be obtained via Eq. (7.23).

Given $p_{SS} = 1/2$ and $p_{FS} = 1/4$, for example, then $P(R_{5,2} = 1) = 0.643$. Further, if $p_{SS} = p_{FS} = p$, $0 < p < 1$ (*i.e.* under the null hypothesis of *i.i.d.*), then $P(R_{5,2} = 1) = 0.4$ for all p, and this result can be easily confirmed by writing out all 32 possible sequences from 5 two-state trials. $\diamond$

Remark 7.1: The Markov chain $\{Y_t\}$ defined above contains three components: (i) the number of successes in the first t trials, (ii) the number of success runs in the first t trials, and (iii) the outcome of the t-th trial (serving as ending block). When t increases, variables (i) and (ii) are non-decreasing. Therefore, when the number of successes n_1 is given, the state space can be reduced by collapsing the states where $N_1 > n_1$ into an absorbing state α. This reduction not only reduces the size of the state space, but also the dimension of the transition probability matrix, and hence computational time. The conditional distribution of R_n given $N_1 = n_1$ will not be affected by this reduction.

Remark 7.2: If the two-state trials are non-homogeneous Markov dependent, the joint distribution of (N_1, R_n) can be obtained in the same way by simply replacing the transition probabilities p_{SS}, p_{SF}, p_{FS}, and p_{FF}, in the transition probability matrix of Eq. (7.21), with time-dependent transition probabilities $p_{SS}(t)$, $p_{SF}(t)$, $p_{FS}(t)$, and $p_{FF}(t)$. Furthermore, if the two-state trials are *i.i.d.* with $p_{SS} = p_{FS} = p$, the conditional distribution of R_n given N_1 under the null hypothesis can be obtained easily. This implies that Theorem 7.2 can be used for computing both critical regions and powers of the runs test.

Critical regions for the test based on success runs, $R_{n,n_1} = (R_n | N_1 = n_1)$, can be constructed using the method of finite Markov chain imbedding. To demonstrate the capabilities of this method, the critical regions of the two-sided success runs test, at the 5% significance level, for $p = 1/2$ and a large sample of $n = 100$ are obtained, and are presented in a graphic as opposed to tabular format in Figure 7.1. Given any number of successes n_1, the hypothesis H_0 can be tested by whether the observed value of R_{n,n_1} is located in the rejection region (**Cr**) or not. As for all tests based on discrete probability distributions, the tail probabilities may not be equal to the assigned significance level. Therefore, Figure 7.1 was constructed based on the closest critical values for both sides.

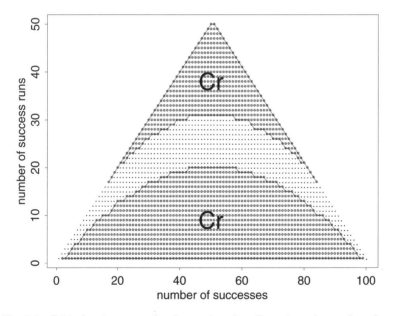

Fig. 7.1 Critical region curves for the test based on R_{100} given the number of successes n_1, with $p = 1/2$ and a significance level of 5%.

Some power functions are plotted in Figure 7.2 for a moderate sample size of $n = 30$, and various sizes of the test. Power comparisons for two-sided tests based on R_{n,n_1} and L_{n,n_1} (the length of the longest success run

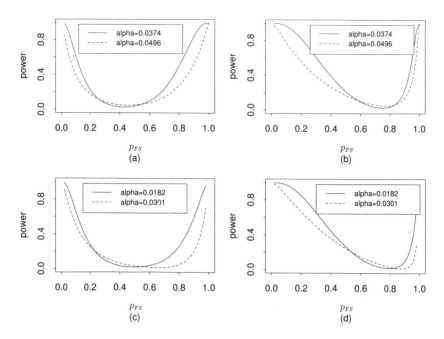

Fig. 7.2   Powers of two-sided tests based on R_{n,n_1} with $n = 30$: (a) $p_{SS} = 0.5, n_1 = n_2$, (b) $p_{SS} = 0.8, n_1 = n_2$, (c) $p_{SS} = 0.5, n_1 = 2/3n_2$, and (d) $p_{SS} = 0.8, n_1 = 2/3n_2$ (alpha = size of the critical region).

given $N_1 = n_1$) are given in Lou (1996).

To illustrate the practical use of the runs test for binary longitudinal data, one subset of data from a multi-center health risk study on asthmatics is analyzed here.

Example 7.4   In Canada, over half a million people suffer from asthma (Wigle 1982), a cause of illness and disability. Despite advances in treatment modalities, the morbidity and mortality associated with asthma have continued to increase in Canada over the past decade. Hence it is vital to have a strategy for the prevention and control of the disease. A multi-center study was conducted by the Gage Research Institute, Toronto, across Canada from November 1991 to May 1993 in order to identify important environmental factors which may increase disease severity or cause acute exacerbation, and to provide insights on regional differences. With the necessary information, a prevention strategy can then be planned, such as better enforcement of environmental controls against certain air pollutants, the administration of vaccines against viral infections, and so on. During the preliminary data analysis, the randomness of daily symptoms was tested. As an example, the occurrence of wheeze, a common discomfort symptom of asthmatics, will be considered here as the outcome variable. It is defined as 'Failure' if there were no symptoms of wheeze on a given day, and as 'Success' otherwise. Of the 59 asthmatics from Vancouver, BC, there were 29 subjects who had both more than two days of no symptoms as well as some symptoms during a 200-day period, and it is these subjects which will be used in the following analysis. The remaining subjects who had symptoms nearly every day or no symptoms at all are not used, as the results are trivial.

As shown in Figure 7.3, there are 22 sequences that have the number of success runs falling inside the critical region curves, and seven sequences which do not. The approximate probabilities for large samples given by Wald and Wolfowitz (1940) were also computed and compared to the exact probabilities, which were obtained easily using the finite Markov chain imbedding approach, and most approximations were found to be significantly inaccurate.

Based on the exact probabilities and Figure 7.3, the null hypothesis of randomness at the 5% significance level is rejected for 22 out of the 29 sequences. Therefore, for about three quarters of the subjects, the occurrence or non-occurrence of wheeze on a given day is somehow related to their condition on previous days. The type of tendency (positive or negative, first- or second-order Markov dependence, *etc.*) can then be modeled.   ◇

7.4 Sequential Continuity

Continuity of care, and in particular provider continuity, is increasingly becoming a central aim of health care policy in a majority of clinical settings. The benefits of broader information exchange between doctor and patient attributed to provider continuity include earlier recognition of health problems and psycho-social effects such as greater patient satisfaction, leading to an overall improvement in the quality of care as well as a reduction in cost.

Continuity of care describes the extent to which information about the diagnosis and management of health problems is conveyed from one visit to the next. This definition, in its broadest sense, includes not only provider

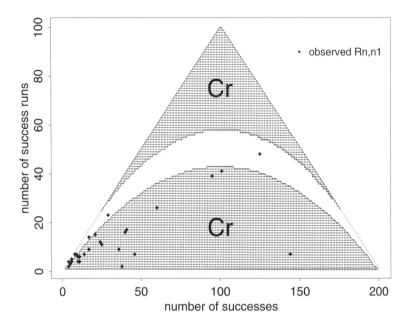

Fig. 7.3 Critical region curves for $n = 200$, and observed numbers of success runs ($\blacklozenge$) for the 29 sequences of Example 7.4.

(*e.g.* physician, health team, HMO) continuity, but also other dimensions such as continuity of medical records and of geographical location of treatment site. Although existing measures for the quantification of continuity can be applied to each of these dimensions in a similar manner, corresponding literature has focused mainly on provider continuity, which is expected to have the greatest impact upon treatment outcome. Over the past two decades, more than a dozen measurement indices of continuity have been proposed, and, in general, they can be categorized into visit-based and individual-based measures. Steinwachs (1979) proposed an individual-based statistic: the sequential continuity measure $SECON$, which is the fraction of sequential visit pairs at which the same provider is seen. Currently it is one of the most commonly used individual-based statistics. A brief introduction to the formulation for the imbedded $SECON$ statistic has been given in Example 2.4. A more complete and detailed mathematical discussion is provided in the following.

Let $\{X_t, t = 0, 1, \cdots\}$ be a stationary sequence of random variables forming a homogeneous Markov chain defined on a finite state space $S = \{1, 2, \cdots, m\}$. Assume the chain is irreducible and aperiodic with transition probability matrix $\boldsymbol{A} = (p_{ij})_{m \times m}$. Let $\boldsymbol{\pi} = (\pi_1, \cdots, \pi_m)$ be the ergodic distribution associated with the Markov chain; *i.e.* for $i = 1, \cdots, m$, $\pi_i = \lim_{n \to \infty} P(X_n = i)$. Define a parameter $\theta = \sum_{i=1}^{m} \pi_i p_{ii}$, which is the sequential continuity measure associated with the Markov chain, and define a sequence of indicator random variables I_t such that $I_t = 1$ if $X_t = X_{t-1}$, $t = 1, 2, \cdots$. Thus $I_t = 1$ if the same health provider is seen at consecutive visits, and zero otherwise, where the random variable X_0 corresponds to the outcome of the initial visit of the patient at time $t = 0$ with initial probability $\boldsymbol{\pi}_0 = (P(X_0 = 1), \cdots, P(X_0 = m))$. Unless specified otherwise, we assume that the initial probability $\boldsymbol{\pi}_0$ is the ergodic distribution $\boldsymbol{\pi}$ of the Markov chain (*i.e.* $\boldsymbol{\pi}_0 = \boldsymbol{\pi}$).

Let $v(T)$, a positive integer random variable defined on $J^+ = \{1, 2, \cdots\}$ with probability mass function $\mu(\cdot)$, be the number of visits, excluding the initial visit, that have occurred up to time T. Steinwachs (1979) introduced the statistic $SECON$ as the outcome average of the sequence $\{I_t\}_{t=1}^{v(T)}$ for a given $v(T) = n$, or in a practical sense, as the fraction of sequential visit pairs at which the same health provider is seen. In our context, $SECON$ estimates θ, with

$$SECON = S_{v(T)}/v(T) \qquad (7.24)$$

and $S_{v(T)} = I_1 + I_2 + \cdots + I_{v(T)}$.

The exact distribution of $SECON$ under the assumption of random assignment (*i.i.d.* with equal probabilities) at each visit was given by Steinwachs (1979). Here we focus on the exact distribution of $SECON$ under the more realistic model of dependence between visits, obtained originally by Fu and Lou (2000a).

If the integer random variable $v(T)$ is independent of the Markov chain $\{X_t\}$, and the initial distribution $\boldsymbol{\pi}_0$ is the ergodic distribution $\boldsymbol{\pi}$ of the Markov chain, then $SECON$ is an unbiased estimator for the sequential continuity parameter θ. This follows from the fact that for every given $v(T) = n$ with $n \in J^+$, $E(SECON|v(T) = n) = E(S_n/n) = \theta$. It is easy to see that the sequence of random variables $\{I_t\}$ induced by the Markov chain $\{X_t\}$ is stationary but not always a Markov chain. Hence the usual Markov chain techniques cannot be applied directly to study the exact distributions of $S_{v(T)}$ and *SECON*. The goal here is to find the exact distributions of $S_{v(T)}$ and *SECON* under one-step Markov dependence between two consecutive events/visits.

Decompose the transition probability matrix A associated with $\{X_t\}$ into two matrices G and D: $A_{m \times m} = (p_{ij})_{m \times m} = G_{m \times m} + D_{m \times m}$, where

$$G_{m \times m} = \begin{pmatrix} 0 & p_{ij} & \cdots \\ & \ddots & \\ \cdots & p_{ij} & 0 \end{pmatrix} \text{ and } D_{m \times m} = \begin{pmatrix} p_{11} & & O \\ & \ddots & \\ O & & p_{mm} \end{pmatrix}. \quad (7.25)$$

Let $\Omega = \{(u, v) : u = 0, 1, \cdots, n, \text{ and } v = 1, 2, \cdots, m\}$ be the state space containing a total of $(n+1)m$ states. Define a homogeneous Markov chain $\{Y_t\}$ on Ω as

$$Y_t = \begin{cases} (\sum_{i=1}^t I_i, X_t) & 1 \leq t \leq n \\ (0, X_0) & t = 0, \end{cases} \quad (7.26)$$

with transition probability matrices, for $t = 1, \cdots, n$, given by

$$M_{(n+1)m \times (n+1)m} = \begin{pmatrix} G & D & O & \cdots & O & O \\ O & G & D & \cdots & O & O \\ & & \ddots & & \ddots & \\ O & & \cdots & & G & D \\ O & & \cdots & & O & I \end{pmatrix}. \quad (7.27)$$

Theorem 7.3 *The distribution of S_n is given by*

$$P(S_n = s) = \boldsymbol{\xi}_0 M^n U'(C_s), \quad 0 \leq s \leq n, \quad (7.28)$$

where $\boldsymbol{\xi}_0 = (\boldsymbol{\pi}_0, \mathbf{0}, \cdots, \mathbf{0}, \cdots, \mathbf{0})_{1 \times (n+1)m}$ is the initial probability vector of Y_0, $\mathbf{0} = (0, \cdots, 0)_{1 \times m}$, $U(C_s) = (0, \cdots, 0, 1, \cdots, 1, 0, \cdots, 0)$ is a $1 \times (n+1)m$ row vector with 1 at the coordinates associated with states in C_s and 0 elsewhere, and the $C_s = [(s, 1), \cdots, (s, m)], s = 0, 1, \cdots, n$, form a partition of Ω. Further, the probability of $S_n = s$ satisfies the following recursive equation:

$$P(S_n = s) = P(S_{n-1} = s) + \sum_{j=1}^m p_{jj} \boldsymbol{\xi}_0 M^{n-1}(e'(s-1, j) - e'(s, j)), \quad (7.29)$$

where $e(s, j) = (0, \cdots, 0, 1, 0, \cdots, 0)_{1 \times (n+1)m}$ is a unit vector associated with the state (s, j).

Proof. Since $\{X_t\}$ is a Markov chain with initial probability $\boldsymbol{\pi}_0$ and transition probability matrix $\boldsymbol{A}$, it follows from Eq. (7.26) that $\{Y_t\}$ is also a Markov chain with initial probability $\boldsymbol{\xi}_0$ and transition probabilities determined by the following: for $0 \leq s \leq n-1$ and $i = 1, \cdots, m$,

$$P(Y_t = (u,v)|Y_{t-1} = (s,i)) = \begin{cases} p_{ii} & \text{if } u = s+1 \text{ and } v = i \\ p_{ij} & \text{if } u = s \text{ and } v = j, \ i \neq j \\ 0 & \text{otherwise.} \end{cases} \quad (7.30)$$

For states with $s = n$, by convention, $P(Y_t = (u,v)|Y_{t-1} = (n,i)) = 1$ if $u = n$ and $v = i$, and 0 otherwise. The transition probability matrix $\boldsymbol{M}$ given in Eq. (7.27) is hence defined. From the definition of the Markov chain $\{Y_t\}$, $P(S_n = s) = P(Y_n \in C_s)$, so Eq. (7.28) is a direct consequence of the Chapman-Kolmogorov equation. The recursive Eq. (7.29) follows immediately from Eq. (7.28) and

$$\boldsymbol{M}\boldsymbol{U}'(C_s) = \boldsymbol{U}'(C_s) + \sum_{j=1}^{m} p_{jj}(\boldsymbol{e}'(s-1,j) - \boldsymbol{e}'(s,j)).$$

$\square$

Note that $\boldsymbol{G}\boldsymbol{D} = \boldsymbol{D}\boldsymbol{G}$ if and only if $p_{11} = p_{22} = \cdots = p_{mm} = \theta > 0$. Given $v(T) = n$, if $\boldsymbol{G}\boldsymbol{D} = \boldsymbol{D}\boldsymbol{G}$, it follows from Eq. (7.28) and

$$\boldsymbol{\xi}_0 \boldsymbol{M}^n = \left[\binom{n}{0} \boldsymbol{\pi}_0 \boldsymbol{G}^n, \binom{n}{1} \boldsymbol{\pi}_0 \boldsymbol{G}^{n-1} \boldsymbol{D}, \cdots, \binom{n}{n} \boldsymbol{\pi}_0 \boldsymbol{D}^n \right] \quad (7.31)$$

that S_n has a binomial distribution

$$P(S_n = s) = \binom{n}{s} \boldsymbol{\pi}_0 \boldsymbol{G}^{n-s} \boldsymbol{D}^s \boldsymbol{1}' = \binom{n}{s} \theta^s (1-\theta)^{n-s}. \quad (7.32)$$

Thus in this special case, the result is independent of the number of available providers m and the initial probability $\boldsymbol{\pi}_0$.

The Markov chain $\{Y_t\}$ defined in Eq. (7.26) can be regarded as a Markov random walk (Pyke 1961). Our construction of $\{Y_t\}$ shows that S_n is finite Markov chain imbeddable. In addition, if $\boldsymbol{G}\boldsymbol{D} = \boldsymbol{D}\boldsymbol{G}$, S_n is finite Markov chain imbeddable of binomial type in the sense of Definition 2.7 (Koutras and Alexandrou 1995).

If $v(T)$ (with probability measure $\mu(\cdot)$ on J^+) is independent of $\{X_t\}$ and has, for example, a truncated Poisson distribution with parameter λ such that $\mu(n) = \lambda^n e^{-\lambda}/(n!(1 - e^{-\lambda}))$, $n \in J^+$, then for cases where

$p_{11} = \cdots = p_{mm} = \theta$, the $SECON$ statistic has the following distribution: for $\alpha \in [0,1]$,

$$P(S_{v(T)}/v(T) \leq \alpha) = \frac{e^{-\lambda}}{1 - e^{-\lambda}} \sum_{n=1}^{\infty} \frac{\lambda^n}{n!} \sum_{s=0}^{[\alpha n]} \binom{n}{s} \theta^s (1 - \theta)^{n-s}. \qquad (7.33)$$

Further, under the above assumptions, the $SECON$ statistic has mean θ and variance

$$Var(S_{v(T)}/v(T)) = \frac{\theta(1 - \theta)e^{-\lambda}}{1 - e^{-\lambda}} \sum_{n=1}^{\infty} \frac{\lambda^n}{n(n!)}. \qquad (7.34)$$

The assumption of independence between the number of visits $v(T)$ and the visit sequence $\{X_t\}$ is crucial for the above results.

Example 7.5   As part of a study of HIV/AIDS patients, we investigated the relationship between continuity of care and overall outcomes, such as total charges and frequency of hospitalization, for patients from the Mount Sinai AIDS Center. This patient population is composed mainly of East and Central Harlem residents. These low-income residential communities with high illicit drug use have among the highest AIDS and tuberculosis prevalences of any community in the country, and it is important to quantify the patterns of care received by patients in such areas.

As an illustration here, we select one set of 45 patients whose primary provider is Medicaid, whose ages are between 25 and 55, and who had regular visits since the initial date of receiving case management care during the period of July 1995 to June 1996. Among the 45 observed sequences, 24 had $SECON$ values equal to one, and the minimum value was 0.3. Hence these patients received much better sequential continuity of care than if they had been randomly assigned to any one of the $m = 13$ providers at each visit, in which case θ would be $1/13$. To gain further insight into the measure θ for these 45 patients, the empirical distribution of $SECON$ is plotted in Figure 7.4, along with one theoretical distribution of $SECON$ where $\theta = 0.8$ and the number of visits $v(T)$ is assumed to follow a Poisson distribution with $\lambda = 8$.

Here, as a reasonable starting point, we naively assumed a symmetric transition probability relationship for the 13-provider Markov chain, resulting in the small discrepancies apparent in Figure 7.4. Note that both distributions are heavily skewed toward the left. A chi-square test with four intervals, each containing at least five observations, was carried out for

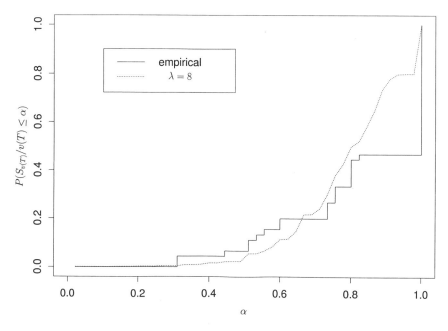

Fig. 7.4   The empirical distribution of $S_{v(T)}/v(T)$ for the 45 patients of Example 7.5, along with the exact distribution of $S_{v(T)}/v(T)$ where $\theta = 0.8$ and $v(T)$ follows a truncated Poisson distribution with $\lambda = 8$.

goodness-of-fit of the theoretical distribution of $SECON$ with parameters $\theta = 0.8$ and $\lambda = 8$, yielding a p-value of 0.09. It's worth mentioning that when the restriction of having at least five observations in each interval is removed, as in using the interval $\alpha \in [0, 0.34]$ which contains two observations, the p-value decreases to less than 0.05. The requirement of at least five observations in each interval was chosen to eliminate potential problems in the chi-square approximation due to extremely small expected values. Based on Figure 7.4 and the chi-square goodness-of-fit test, it appears that the theoretical distribution approximates the empirical distribution well in the center, and less well in the two tails. With proper estimation of the transition probabilities, we can then model the relationship between the distribution of $S_{v(T)}/v(T)$ and the characteristics of the patient population (health care provider, race, *etc.*) and further with overall outcomes such as total costs and quality of care.   ◇

7.5   Quality Control Schemes

Basic control schemes such as Cumulative Sum (CUSUM), Exponentially Weighted Moving Average (EWMA), and the Shewhart chart have found widespread application in improving the quality of manufactured goods and services. It is known that there is no single control scheme that is optimal in detecting all types of changes and shifts. Recently, multiple control schemes (compound sets of control rules) have been used for monitoring various manufacturing process parameters, such as the mean, variance, and proportion. One example of a multiple control scheme is the Western Electric Control Scheme (see Montgomery 2001, p. 176):

(1) One or more points outside of the control limit,
(2) Two of three consecutive points outside the two-sigma warning limits but still inside the control limits,
(3) Four of five consecutive points beyond the one-sigma limits, and
(4) A run of eight consecutive points on one side of the center line.

A number of methods have been used to investigate the run-length distributions (waiting-time distributions for an out-of-control signal) and the average run lengths (ARLs) for various control schemes; a basic overview of the area may be gained from the following references: Barnard (1959), Ewan and Kemp (1960), Brook and Evans (1972), Lucas and Crosier (1982), and Chao (1999). In this section, we adopt a general method based on discretization and the finite Markov chain imbedding technique to study the distributions and ARLs of various control schemes, as given by Fu, Spiring and Xie (2002). The reasons for imposing discretization are two-fold: (i) to allow for the efficient application of the method of finite Markov chain imbedding in computing the run-length distribution when the control scheme $\{S_n\}$ is a sequence of continuous random variables, and (ii) to avoid the mathematical complexity of other methods in computing the ARL for multiple control schemes (see Montgomery 2001). Technically speaking, we discretize the range space into m states plus control limits (absorbing states), and then the quality control scheme can be viewed as the limiting control scheme based on a sequence of multi-state trials, $\{S_n(m)\}$, as $m \to \infty$. In plain words, for each control scheme $\{S_n\}$, we construct a sequence of control schemes $\{S_n(m)\}$ defined on a finite state space. Under very mild assumptions, we expect the discretized control schemes to perform as $\{S_n\}$, in the limit $m \to \infty$.

7.5.1 *Simple Control Schemes*

Let $\{Z_n\}$ be a sequence of *i.i.d.* continuous random variables. The simple control schemes known as CUSUM, EWMA and the Shewhart chart are perhaps the three most commonly used in practice. Mathematically, these schemes can be described as follows.

(1) CUSUM control scheme: Given some upper control limit (UCL), $h > 0$, the upper-sided CUSUM is defined as

$$S_0 = 0, S_n = \max\{0, S_{n-1} + Z_n\}, \text{ for } n = 1, 2, \cdots. \tag{7.35}$$

The process being monitored is out-of-control at stage n (after n trials) if $S_i < h$ for all $i = 0, 1, 2, \cdots, n-1$, and $S_n \geq h$. The main use of the upper-sided CUSUM control scheme is to detect a small upward shift. Similarly, we can define the lower-sided CUSUM as

$$S_0 = 0, S_n = \min\{0, S_{n-1} + Z_n\}, \text{ for } n = 1, 2, \cdots. \tag{7.36}$$

The process is out-of-control at stage n if $S_i > -h$ for all $i = 0, 1, \cdots, n-1$, and $S_n \leq -h$, where $-h$ is referred to as the lower control limit (LCL).

(2) EWMA control scheme: For given $0 < \lambda \leq 1$ and $h > 0$, the EWMA scheme can be defined as

$$S_0 = 0, S_n = (1 - \lambda)S_{n-1} + \lambda Z_n, \text{ for } n = 1, 2, \cdots. \tag{7.37}$$

The process is out-of-control at stage n if $-h < S_i < h$ for all $i = 0, 1, \cdots, n-1$, and $S_n \leq -h$ or $S_n \geq h$.

(3) Shewhart control scheme: The Shewhart control scheme is the special case of the EWMA control scheme with $\lambda = 1$:

$$S_0 = 0, S_n = Z_n, \text{ for } n = 1, 2, \cdots.$$

In general, the upper and lower control limits are functions of n. For example, in practice, for the two-sided CUSUM control scheme a V-mask is often used, for the EWMA control scheme linear control limits are sometimes applied, but for the Shewhart control scheme the upper and lower limits are typically constant. For simplicity in presenting the application of the finite Markov chain imbedding technique to the analysis of quality control schemes, we now focus our discussion on the upper-sided CUSUM control scheme (denoted hereafter simply by S_n), and choose a constant control limit ($h > 0$). At the end of this subsection, we briefly indicate how to extend the methodology to non-constant control limits. However, the theoretical results developed in this section for the upper-sided CUSUM control scheme also hold for the EWMA and Shewhart schemes.

Given a constant control limit $h > 0$ and $S_0 = 0$ with probability one, define

$$W = \inf\{n \geq 1 : S_n \geq h | S_0 = 0\}$$

as the run-length (waiting-time or hitting-time) random variable induced by the upper-sided CUSUM control chart.

Let $f(z)$ be a continuous common density function for a sequence of i.i.d. random variables $\{Z_n\}$. Given the upper control limit $h > 0$ and a positive integer m, let $\{S_n(m)\}$ be the sequence of discretized random variables of the sequence $\{S_n\}$, where $S_n(m)$ takes values $a_0 = 0$, $a_i = id$, for $i = 1, \cdots, m$, and $a_{m+1} = h$ with $d = h/(m+1)$. We define a discretized upper-sided CUSUM as follows: $S_0(m) = a_0 = 0$ and

$$S_n(m) = \begin{cases} a_0 & \text{if } 0 \leq \max\{0, S_{n-1}(m) + Z_n\} < a_0 + \frac{d}{2} \\ a_j & \text{if } a_{j-1} + \frac{d}{2} \leq \max\{0, S_{n-1}(m) + Z_n\} < a_j + \frac{d}{2}, \\ & \quad \text{for } j = 1, \cdots, m \\ a_{m+1} & \text{if } a_m + \frac{d}{2} \leq \max\{0, S_{n-1}(m) + Z_n\}, \end{cases}$$

(7.38)

for $n = 1, 2, \cdots$. It is easy to see from our construction that $\{S_n(m)\}$ forms a homogeneous Markov chain defined on the state space $\Omega = \{a_0, a_1, \cdots, a_m, a_{m+1}\}$ with transition probability matrix

$$\boldsymbol{M} = \begin{pmatrix} p_{00} & p_{01} & \cdots & p_{0m} & p_{0(m+1)} \\ p_{10} & p_{11} & \cdots & p_{1m} & p_{1(m+1)} \\ \vdots & \vdots & \ddots & \vdots & \vdots \\ p_{m0} & p_{m1} & \cdots & p_{mm} & p_{m(m+1)} \\ 0 & 0 & \cdots & 0 & 1 \end{pmatrix} = \left(\begin{array}{c|c} \boldsymbol{A}(m) & \boldsymbol{B}(m) \\ \hline \boldsymbol{0} & 1 \end{array} \right),$$

(7.39)

where the state a_{m+1} is an absorbing state, $\boldsymbol{A}(m)$ is the $(m+1) \times (m+1)$ essential submatrix, $\boldsymbol{B}(m)$ is an $(m+1) \times 1$ vector,

$$p_{ij} = \int_{(a_{j-1}-a_i)+\frac{d}{2}}^{(a_j-a_i)+\frac{d}{2}} f(z)dz, \quad i = 0, 1, 2, \cdots, m, \ j = 1, 2, \cdots, m,$$

$$p_{i0} = \int_{-\infty}^{(a_0-a_i)+\frac{d}{2}} f(z)dz, \quad i = 0, 1, 2, \cdots, m,$$

$$p_{i(m+1)} = \int_{(a_m-a_i)+\frac{d}{2}}^{\infty} f(z)dz, \quad i = 0, 1, 2, \cdots, m,$$

$$p_{(m+1)(m+1)} = 1 \text{ (stay "out of control"), and} \tag{7.40}$$

$$p_{(m+1)j} = 0, \quad j = 0, 1, 2, \cdots, m \text{ (cannot return once beyond UCL)}.$$

Note that, for every $i = 0, 1, \cdots, m$,

$$\sum_{j=0}^{m+1} p_{ij} = \int_{-\infty}^{(a_0-a_i)+\frac{d}{2}} f(z)dz + \sum_{j=1}^{m} \int_{(a_{j-1}-a_i)+\frac{d}{2}}^{(a_j-a_i)+\frac{d}{2}} f(z)dz$$

$$+ \int_{(a_m-a_i)+\frac{d}{2}}^{\infty} f(z)dz \equiv 1.$$

Hence, the transition probability matrix $\boldsymbol{M}$ associated with the upper-sided CUSUM control scheme consists of elements p_{ij}, $i, j = 0, 1, \cdots, m+1$, as defined in Eq. (7.40), where each p_{ij} represents the probability of the control statistic moving from state i to state j.

Let the random variable of run length induced by the finite Markov chain $\{S_n(m)\}$ be $W(m)$, such that

$$W(m) = \inf\{n \geq 1 : S_n(m) = a_{m+1}|S_0(m) = a_0\}, \tag{7.41}$$

where $a_0 = 0$. Throughout the following, we assume the initial distribution $P(S_0(m) = S_0 = a_0) \equiv 1$; *i.e.* $\boldsymbol{\xi}(m) = (1, 0, \cdots, 0)$ and $\boldsymbol{\xi}_0 = (\boldsymbol{\xi}(m) : 0)$. The run-length distribution and its mean and variance are given by the following results.

Theorem 7.4 *Given m, h, and $\boldsymbol{\xi}(m)$, we have*

$$(i) \quad P(W(m) = n|\boldsymbol{\xi}(m)) = \boldsymbol{\xi}(m)\boldsymbol{A}^{n-1}(m)(\boldsymbol{I} - \boldsymbol{A}(m))\boldsymbol{1}',$$

$$(ii) \quad E[\exp(sW(m))] = 1 + (e^s - 1)\boldsymbol{\xi}(m)(\boldsymbol{I} - e^s\boldsymbol{A}(m))^{-1}\boldsymbol{1}',$$

$$(iii) \quad E[W(m)] = \boldsymbol{\xi}(m)(\boldsymbol{I} - \boldsymbol{A}(m))^{-1}\boldsymbol{1}', \text{ and}$$

$$(iv) \quad E[W^2(m)] = \boldsymbol{\xi}(m)(\boldsymbol{I} + \boldsymbol{A}(m))(\boldsymbol{I} - \boldsymbol{A}(m))^{-2}\boldsymbol{1}'.$$

Proof. The result (i) follows immediately from Eqs. (7.39) and (7.41), and from Theorem 5.1. With straightforward manipulation, the result (ii) follows from the definition of the moment generating function. The results (iii) and (iv) are a direct consequence of

$$E[W^i(m)] = \frac{d^i}{ds^i} E[\exp(sW(m))]|_{s=0}, \quad i = 1, 2.$$

$\square$

Theorem 7.5 *For every given n,*

$$(i) \lim_{m \to \infty} P(W(m) > n | S_0(m) = a_0) = P(W > n | S_0 = a_0), \text{ and}$$

$$(ii) \lim_{m \to \infty} P(W(m) > n | S_0(m) = a_0) = p_{00} P(W > n - 1 | S_1 = a_0)$$

$$+ \int_0^h P(W > n - 1 | S_1 = z) f(z) dz.$$

Proof.   Given m, it follows from the definitions of W and $W(m)$ that for every $n = 1, 2, \cdots$, we have

$$P(W > n | S_0 = a_0) = P(S_n < h | S_0 = a_0), \tag{7.42}$$

and

$$P(W(m) > n | S_0(m) = a_0) = P(S_n(m) < h | S_0(m) = a_0).$$

It follows from the construction of p_{ij} and the definition of $S_n(m)$ that, for all n, $S_n(m)$ converges to S_n as $m \to \infty$; *i.e.*

$$\lim_{m \to \infty} P(S_n(m) < h | S_0(m) = a_0) = P(S_n < h | S_0 = a_0). \tag{7.43}$$

The result (i) follows immediately from Eq. (7.43). Further, since $[0, h]$ is a compact set, the convergence (for fixed n) is uniform. It follows from Theorem 5.1, the definition of integration, and Eq. (7.41), that as $m \to \infty$ $(d \to 0)$,

$$P(W(m) > n | S_0(m) = a_0) = \sum_{i=0}^m p_{0i} e_i \mathbf{A}^{n-1}(m) \mathbf{1}'(m)$$

$$= p_{00} P(W(m) > n - 1 | S_1(m) = a_0) + \sum_{i=1}^m p_{0i} P(W(m) > n - 1 | S_1(m) = a_i)$$

$$\to p_{00} P(W > n - 1 | S_1 = a_0) + \int_0^h P(W > n - 1 | S_1 = z) f(z) dz.$$

$\square$

To illustrate the methodology developed above, two examples with different probability densities are given in the following to provide some numerical results of ARLs and standard deviations for the upper-sided CUSUM, the EWMA, and the Shewhart control schemes.

Example 7.6   (Upper-sided CUSUM). Consider that (i) UCL $= h = 3$, (ii) $m = 20, 40, \cdots, 1000$, and (iii) the sequence of random variable $\{Z_n\}$

has a common exponential density function $f(z) = \exp\{-z\}$, $0 < z < \infty$. The transition probabilities of $A(m)$ are obtained from Eq. (7.40), and Table 7.4 gives the expected mean and standard deviation for $W(m)$ using Theorem 7.4.

Table 7.4 Expected mean and standard deviation of run length for the upper-sided CUSUM control scheme in Example 7.6.

m	ARL= $E[W(m)]$	$\sqrt{Var[W(m)]}$
20	3.78584	1.84553
40	3.89026	1.79263
80	3.94445	1.76339
200	3.97761	1.74486
500	3.99101	1.73723
1000	3.99550	1.73466

In view of Table 7.4, on the average, after four trials the process is "out of control". This demonstrates that CUSUM detects an upward shift very quickly, due to the fact that the process $S_n = \max\{0, S_{n-1} + Z_n\}$ is monotonically increasing, with an average upshift in this example of $EZ_n = 1$ at every step. This property can be viewed as an advantage, and sometimes as a disadvantage, of using the CUSUM control scheme. ◇

Example 7.7 (EWMA control scheme). We assume (i) the UCL and LCL (Upper and Lower Control Limits) are h and $-h$, respectively, (ii) each side of $a_0 = 0$ has m states and an absorbing state $a_{m+1} = h$ (or $-h$) resulting in $2(m+1)+1$ states in total, (iii) states are equally spaced with $d = h/(m+1)$, and (iv) a standard normal density function

$$f(z) = \frac{1}{\sqrt{2\pi}}e^{-z^2/2}, \quad -\infty < z < \infty.$$

Given $S_{n-1}(m) = id$ and $S_n(m) = jd$, it follows from the definition of the EWMA scheme given in Eq. (7.37) that z has to satisfy the following two inequalities:

$$(1-\lambda)id + \lambda z \geq (j - \frac{1}{2})d, \quad \text{and}$$
$$(1-\lambda)id + \lambda z \leq (j + \frac{1}{2})d.$$

Hence the transition probabilities of $\boldsymbol{A}(m)$ are, for $i, j = -m, \cdots, -1, 0, 1,$ $\cdots, m$:

$$p_{ij} = \Phi\left[\frac{1}{\lambda}\{(j + \frac{1}{2})d - (1 - \lambda)id\}\right] - \Phi\left[\frac{1}{\lambda}\{(j - \frac{1}{2})d - (1 - \lambda)id\}\right],$$

$$p_{i(m+1)} = 1 - \Phi\left[\frac{1}{\lambda}\{(m + \frac{1}{2})d - (1 - \lambda)id\}\right],$$

$$p_{i(-m-1)} = \Phi\left[\frac{1}{\lambda}\{(-m - \frac{1}{2})d - (1 - \lambda)id\}\right],$$

$$p_{(m+1)j} = p_{(-m-1)j} \equiv 0, \text{ and}$$

$$p_{(m+1)(m+1)} = p_{(-m-1)(-m-1)} \equiv 1, \qquad\qquad (7.44)$$

where each p_{ij} represents the probability of the control statistic moving from state i to state j, and $\Phi(\cdot)$ is the standard cumulative normal distribution function.

Consider $h = 3, m = 7$, and $\lambda = 0.2$. For the EWMA scheme based on the standard normal distribution, Table 7.5 gives the run-length mean and standard deviation holding all but the number of divisions (states) constant, using the results from Theorem 7.4(iii).

Table 7.5 Expected mean and standard deviation of run length for the EWMA control scheme in Example 7.7.

m	ARL= $E[W(m)]$	$\sqrt{Var[W(m)]}$
12	514.266	509.678
22	545.073	540.540
32	552.691	548.172
52	557.075	552.565
102	559.101	554.594
502	559.636	555.132

When λ is set to 1, the EWMA control scheme simplifies to the well-known Shewhart scheme. For a Shewhart chart, the ARL and the variance of the run length are $1/p$ and $1/p^2$, respectively, where p is the probability of an out-of-control signal detected by the Shewhart control scheme:

$$p = 2\int_3^\infty \frac{1}{\sqrt{2\pi}} e^{-\frac{1}{2}z^2} dz \cong 0.00269980.$$

For $m = 502$ in our imbedding framework, it can be seen from Table 7.6 for the Shewhart chart that the ARL and the standard deviation of $W(m)$ are equal to 370.377 and 369.877, respectively. The numerical results converge to the theoretical results (ARL=standard deviation=$1/p = 370.398$) as m increases $(m \to \infty)$. ◇

Table 7.6 Expected mean and standard deviation of run length for the Shewhart control scheme in Example 7.7.

m	ARL= $E[W(m)]$	$\sqrt{Var[W(m)]}$
12	345.711	345.210
22	362.443	361.943
32	366.551	366.051
52	368.918	368.418
102	370.004	369.504
502	370.377	369.877

Note that the procedure developed here is independent of the distribution associated with the characteristics being monitored, and requires only the density function (*i.e.* $f(z)$) to be specified in order to determine the transition probability matrix. This allows more flexibility in the selection of density functions for proper modeling of the sequence $\{Z_n\}$ under realistic conditions.

With some modifications, the method can also be extended to the case where the control limit h is a function of n. Then the imbedded chain is a non-homogeneous, tree-structured Markov chain (*cf.* Section 2.3). The state spaces and the transition probability matrices need to be constructed for every $n = 1, 2, \cdots$.

7.5.2 *Compound Control Schemes*

A control scheme composed of two or more simple control schemes $\phi_1, \cdots, \phi_l$ $(l \geq 2)$ is referred to as a compound control scheme $\phi = \cup_{i=1}^{l} \phi_i$ in the sense that an "out-of-control" signal is issued if and only if any one of the simple control schemes has issued an "out-of-control". This is equivalent to saying that the run length (waiting time) of a compound control scheme ϕ can be defined as

$$W(\phi) = \min(W(\phi_1), \cdots, W(\phi_l)), \tag{7.45}$$

where $W(\phi_i)$ is the out-of-control run length for the simple control scheme ϕ_i. The technique for obtaining the distribution of the waiting time of a compound pattern developed in Chapter 5 has been extended recently by Fu, Shmueli and Chang (2002) to study the compound control scheme. To illustrate this extension, we give the following example.

Example 7.8 Let's consider an upper-sided CUSUM control scheme, which, in addition to the upper control limit h, also has a "warning limit" h^*. The compound control scheme $\phi = \phi_1 \cup \phi_2$ is to signal at time t if the scheme ϕ_1, the CUSUM control scheme $S_t = \max\{0, S_{t-1} + X_t\}$, exceeds the upper control limit at time t, or if the scheme ϕ_2 signals that two of three consecutive CUSUM statistics fall in the interval between the warning and control limits, $[h^*, h)$. We assume that the process outcomes X_t are i.i.d. $N(0, 1)$.

Let $D(X)$ be the discretized standard normal variable X in the following sense,

$$D(X) = i\Delta, \quad i = 0, \pm 1, \pm 2, \cdots, \pm(m+1), \tag{7.46}$$

where Δ is a small positive constant $(\Delta > 0)$.

We define $p_i = P(D(X) = i\Delta)$ and $F(i) = P(D(X) \le i\Delta)$ as follows:

$$p_i = \int_{(i-0.5)\Delta}^{(i+0.5)\Delta} \frac{1}{\sqrt{2\pi}} e^{-1/2x^2} dx, \quad i = 0, \pm 1, \cdots, \pm m,$$

$$p_{(m+1)} = \int_{[(m+1)-0.5]\Delta}^{\infty} \frac{1}{\sqrt{2\pi}} e^{-1/2x^2} dx, \tag{7.47}$$

$$p_{-(m+1)} = \int_{-\infty}^{[-(m+1)+0.5]\Delta} \frac{1}{\sqrt{2\pi}} e^{-1/2x^2} dx, \quad \text{and}$$

$$F(i) = \sum_{j=-(m+1)}^{i} p_j.$$

We denote the upper control limit by $h = (m+1)\Delta$ and the warning limit by $h^* = m^*\Delta$, $m^* < m$. Then S_t takes on the values $i\Delta$, $i = 0, 1, \cdots, (m+1)$, in the interval $[0, h]$.

In addition, the chart in this example is naturally divided by the second rule ϕ_2 into three regions: for $t \geq 1$,

$$R(S_t) = \begin{cases} r_1 & \text{if } 0 \leq S_t < h^* \\ r_2 & \text{if } h^* \leq S_t < h \\ r_3 & \text{if } S_t \geq h, \end{cases} \quad (7.48)$$

and we define an initial state $R(S_0) = \emptyset$.

For simplicity, we write $S_t = i$ to denote $S_t = i\Delta$. Define a state space

$$\Omega = \{(\emptyset, \emptyset), (\emptyset, 0), \cdots, (\emptyset, m), (r_1, 0), \cdots, (r_1, m),$$
$$(r_2, 0), \cdots, (r_2, m^* - 1), \alpha\}, \quad (7.49)$$

with $2 + 2(m + 1) + m^*$ states. States that include r_3 or $(m + 1)$ as one of the coordinates can be combined into the absorbing state α. Using the forward and backward principle, we define a homogeneous Markov chain $\{Y_t\}$ on the state space Ω as, for $t \geq 1$,

$$Y_t(X_1, \cdots, X_t : \phi_1, \phi_2) = \begin{cases} \alpha & \text{if either } W(\phi_1) \leq t \\ & \text{or } W(\phi_2) \leq t \\ [R(S_{t-1}), S_t] & \text{otherwise.} \end{cases} \quad (7.50)$$

Note that given $Y_t = [R(S_{t-1}), S_t]$ at time t, it means, by the definition of Y_t, that, for every $1 \leq j \leq t$, $R(S_{j-1})$ and S_j have to satisfy two conditions: (i) $S_j < (m + 1)\Delta$ and (ii) $R(S_{j-1})$ and $R(S_j)$ can't both be equal to r_2. Further, for $t \geq 1$, note that if Y_t is *not* in the absorbing state, it will *not* move into the absorbing state α at time $t + 1$ if (a) $\{S_{t+1} < h^*\}$, or if (b) $\{h^* \leq S_{t+1} < h, R(S_{t-1}) = \emptyset \text{ and } R(S_t) = r_1\}$, or if (c) $\{h^* \leq S_{t+1} < h, R(S_{t-1}) = R(S_t) = r_1\}$. It follows from Eqs. (7.47) and (7.50) that the transition probabilities of the imbedded Markov chain can be specified by the following equation: for $t \geq 1$,

$$P(Y_{t+1} = [R(S_t), S_{t+1}]|Y_t = [R(S_{t-1}), S_t]) \quad (7.51)$$
$$= \begin{cases} F(X_{t+1} = -S_t) & \text{if } S_{t+1} = 0 \\ P(X_{t+1} = S_{t+1} - S_t) & \text{if } 0 < S_{t+1} < h^* \\ P(X_{t+1} = S_{t+1} - S_t) & \text{if condition (b) or (c) holds} \\ 0 & \text{otherwise.} \end{cases}$$

Since the transition probabilities of each row add to 1, the probability of moving into the absorbing state α can be obtained by subtracting all the other probabilities in the row from 1: *i.e.*,

$$P(Y_{t+1} = \alpha | Y_t = [R(S_{t-1}), S_t]) \tag{7.52}$$
$$= 1 - \sum_{[R(S_t), S_{t+1}] \in \Omega - \alpha} P(Y_{t+1} = [R(S_t), S_{t+1}] | Y_t = [R(S_{t-1}), S_t]).$$

To illustrate such a transition probability matrix, we choose $m = 3$, $h = 4$ ($\Delta = h/(m+1) = 1$) and $h^* = 2$ ($m^* = h^*/\Delta = 2$). Then, $D(X_t) = i\Delta = i$, $i = 0, \pm1, \pm2, \pm3, \pm4$, and S_t can take on the values $i = 0, 1, 2, 3, 4$ (with 4 denoting the absorbing state α). The transition probability matrix M is given by

$$
\begin{array}{c}
(\emptyset,\emptyset) \\
(\emptyset,0) \\
(\emptyset,1) \\
(\emptyset,2) \\
(\emptyset,3) \\
(r_1,0) \\
(r_1,1) \\
(r_1,2) \\
(r_1,3) \\
(r_2,0) \\
(r_2,1) \\
\alpha
\end{array}
\left(
\begin{array}{ccccccccccc|c}
0 & F(0) & p_1 & p_2 & p_3 & 0 & 0 & 0 & 0 & 0 & 0 & 1-F(3) \\
0 & 0 & 0 & 0 & 0 & F(0) & p_1 & p_2 & p_3 & 0 & 0 & 1-F(3) \\
0 & 0 & 0 & 0 & 0 & F(-1) & p_0 & p_1 & p_2 & 0 & 0 & 1-F(2) \\
0 & 0 & 0 & 0 & 0 & 0 & 0 & 0 & 0 & F(-2) & p_{-1} & 1-F(-1) \\
0 & 0 & 0 & 0 & 0 & 0 & 0 & 0 & 0 & F(-3) & p_{-2} & 1-F(-2) \\
0 & 0 & 0 & 0 & 0 & F(0) & p_1 & p_2 & p_3 & 0 & 0 & 1-F(3) \\
0 & 0 & 0 & 0 & 0 & F(-1) & p_0 & p_1 & p_2 & 0 & 0 & 1-F(2) \\
0 & 0 & 0 & 0 & 0 & 0 & 0 & 0 & 0 & F(-2) & p_{-1} & 1-F(-1) \\
0 & 0 & 0 & 0 & 0 & 0 & 0 & 0 & 0 & F(-3) & p_{-2} & 1-F(-2) \\
0 & 0 & 0 & 0 & 0 & F(0) & p_1 & 0 & 0 & 0 & 0 & 1-F(1) \\
0 & 0 & 0 & 0 & 0 & F(-1) & p_0 & 0 & 0 & 0 & 0 & 1-F(0) \\
0 & 0 & 0 & 0 & 0 & 0 & 0 & 0 & 0 & 0 & 0 & 1
\end{array}
\right)
$$

and it is again written in the form

$$M = \left(\begin{array}{c|c} N & C \\ \hline 0 & 1 \end{array} \right).$$

Note that here we assume that S_0 is at the initial state $\emptyset$ with probability one: *i.e.* $P(S_0 = \emptyset) \equiv 1$. Hence the states $(\emptyset,\emptyset)$, $(\emptyset,0)$, $\cdots$, $(\emptyset,m)$ are needed so that Y_0 and Y_1 of the imbedded Markov chain $\{Y_t\}$ can be properly defined with the initial distribution $P(Y_0 = (\emptyset,\emptyset)) \equiv 1$.

Traditionally, the mean and standard deviation of the run-length distribution are used to compare the performance of control schemes. Since our numerical results show that the run-length distribution for a compound control scheme is rather skewed toward the right, we feel that displaying the quartiles along with the mean and standard deviation of the distribution is more proper, as in Table 7.7, for example. In general, the run-length distribution of a compound control scheme becomes more skewed as the number of simple control schemes involved increases. It can be seen from

Table 7.7 Mean, standard deviation and quartiles of run length in Example 7.8.

m	ARL= $E[W(m)]$	$\sqrt{Var[W(m)]}$	Q_1	Q_2	Q_3
5	11.739	9.386	4.626	8.444	14.904
14	12.749	10.187	5.066	9.213	16.232
29	13.103	10.473	5.223	9.486	16.701
74	13.319	10.649	5.316	9.649	16.976
149	13.392	10.709	5.348	9.703	17.075
299	13.428	10.738	5.364	9.730	17.124
749	13.450	10.756	5.373	9.747	17.154
1499	13.457	10.762	5.376	9.753	17.164
1874	13.459	10.764	5.376	9.753	17.164

Table 7.7 that the numerical results are already rather accurate for the case of $m = 299$, and converge fast.   ◇

Generally speaking, so far there are no analytical formulae for the ARLs of compound control schemes (see Montgomery 2001), not even for the *i.i.d.* cases. The finite Markov chain imbedding technique always provides the numerical solutions for the distributions of run length (including ARL, quantiles, variance, *etc.*) for both simple and compound control schemes.

Bibliography

Abramson, M. and Moser, W. O. J. (1967). Permutations without rising or falling w-sequences. *Annals of Mathematical Statistics* **38**, 1245–1254.

Aki, S. (1985). Discrete distributions of order k on a binary sequence. *Annals of the Institute of Statistical Mathematics* **37**, 205–224.

Aki, S. (1997). On sooner and later problems between success and failure runs. *Advances in Combinatorial Methods and Applications to Probability and Statistics* (ed. N. Balakrishnan), Birkhäuser, Boston, 385–400.

Aki, S. (1999). Distributions of runs and consecutive systems on directed trees. *Annals of the Institute of Statistical Mathematics* **51**, 1–15.

Aki, S., Balakrishnan, N. and Mohanty, S. G. (1996). Sooner and later waiting time problems for success and failure runs in higher order Markov dependent trials. *Annals of the Institute of Statistical Mathematics* **48**, 773–787.

Aki, S. and Hirano, K. (1988). Some characteristics of the binomial distribution of order k and related distributions. *Statistical Theory and Data Analysis* II (ed. K. Matusita), North-Holland, Amsterdam, 211–222.

Aki, S. and Hirano K. (1999). Sooner and later waiting time problems for runs in Markov dependent bivariate trials. *Annals of the Institute of Statistical Mathematics* **51**, 17–29.

Aki, S. and Hirano K. (2000). Numbers of success-runs of specified length until certain stopping time rules and generalized binomial distributions of order k. *Annals of the Institute of Statistical Mathematics* **52**, 767–777.

Aki, S., Kuboki, H. and Hirano, K. (1984). On discrete distributions of order k. *Annals of the Institute of Statistical Mathematics* **36**, 431–440.

Antzoulakos, D. L. (1999). On waiting time problems associated with runs in Markov dependent trials. *Annals of the Institute of Statistical Mathematics* **51**, 323–330.

Antzoulakos, D. L. (2001). Waiting times for patterns in a sequence of multistate trials. *Journal of Applied Probability* **38**, 508–518.

Balakrishnan, N. and Koutras, M. V. (2002). *Runs and Scans with Applications*,

Wiley, New York.

Balasubramanian, K., Viveros, R. and Balakrishnan, N. (1993). Sooner and later waiting time problems for Markovian Bernoulli trials. *Statistics and Probability Letters* **18**, 153–161.

Barnard, G. A. (1959). Control charts and stochastic processes. *Journal of the Royal Statistical Society, Series B* **21**, 239–271.

Barton, D. E. and David, F. N. (1958). Non-randomness in a sequence of two alternatives: II. Runs test. *Biometrika* **45**, 253–256.

Bateman, G. (1948). On the power function of the longest run as a test for randomness in a sequence of alternatives. *Biometrika* **35**, 97–112.

Boutsikas, M. V. and Koutras, M. V. (2000a). Generalized reliability bounds for coherent structures. *Journal of Applied Probability* **37**, 778–794.

Boutsikas, M. V. and Koutras, M. V. (2000b). Reliability approximation for Markov chain imbeddable systems. *Methodology and Computing in Applied Probability* **2**, 393–411.

Brook, D. and Evans, D. A. (1972). An approach to the probability distribution of cusum run length. *Biometrika* **59**, 539–549.

Cai, J. (1994). Reliability of a large consecutive-k-out-of-r-from-n:F system with unequal component-reliability. *IEEE Transactions on Reliability* **43**, 107–111.

Carlitz, L. (1964). Extended Bernoulli and Eulerian numbers. *Duke Mathematical Journal* **31**, 667–689.

Chao, M. T. (1999). Applications of Markov chains in quality-related matters. *Statistical Process Monitoring and Optimization* (eds. S. H. Park and G. G. Vining), Marcel Dekker, New York, 175–188.

Chao, M. T. and Fu, J. C. (1989). A limit theorem of certain repairable systems. *Annals of the Institute of Statistical Mathematics* **41**, 809–818.

Chao, M. T. and Fu, J. C. (1991). The reliability of large series system under a Markovian structure. *Advances in Applied Probability* **23**, 894–908.

Chao, M. T., Fu, J. C. and Koutras, M. V. (1995). Survey of reliability studies of consecutive-k-out-of-n:F and related systems. *IEEE Transactions on Reliability* **44**, 120–127.

Chao, M. T. and Lin, G. D. (1984). Economical design of large consecutive-k-out-of-n:F systems. *IEEE Transactions on Reliability* **33**, 411–413.

Chen, J. and Glaz, J. (1997). Approximations and inequalities for the distribution of a scan statistic for 0-1 Bernoulli trials. *Advances in the Theory and Practice of Statistics* (eds. N. L. Johnson and N. Balakrishnan), Wiley, New York, 285–298.

Chen, J. and Glaz, J. (1999). Approximations for the distribution and the moments of discrete scan statistics. *Scan Statistics and Applications* (eds. J. Glaz and N. Balakrishnan), Birkhäuser, Boston, 27–66.

Cheung, L. K. W. (2002). *Statistical Pattern Recognition in Genomic DNA Sequences*. Ph.D. Dissertation, Department of Statistics, University of Manitoba, Canada.

Chiang, D. T. and Niu, S. C. (1981). Reliability of consecutive-k-out-of-n:F systems. *IEEE Transactions on Reliability* **30**, 87–89.

Chrysaphinou, O. and Papastavridis, S. (1988). A limit theorem on the number of overlapping appearances of a pattern in a sequence of independent trials. *Probability Theory and Related Fields* **79**, 129–143.

Cochran, W. G. (1938). An extension of Gold's method for examining the apparent persistence of one type of weather. *Quarterly Journal of the Royal Meteorological Society* **64**, 631–634.

Csörgö, S. (1979). Erdös-Rényi laws. *Annals of Statistics* **7**, 772–787.

David, F. N. (1947). A power function for tests of randomness in a sequence of alternatives. *Biometrika* **34**, 335–339.

David, F. N. and Barton, D. E. (1962). *Combinatorial Chance*, Hafner, New York.

Derman, G., Lieberman, G. J. and Ross, S. M. (1982). On the consecutive-k-out-of-n:F system. *IEEE Transactions on Reliability* **31**, 57–63.

Dillon, J. F. and Roselle, D. P. (1969). Simon Newcomb's problem. *SIAM Journal on Applied Mathematics* **17**, 1086–1093.

Doi, M. and Yamamoto, E. (1998). On the joint distribution of runs in a sequence of multi-state trials. *Statistics and Probability Letters* **39**, 133–141.

Dwass, M. (1973). The number of increases in a random permutation. *Journal of Combinatorial Theory, Series A* **15**, 192–199.

Ebneshahrashoob, M. and Sobel, M. (1990). Sooner and later problems for Bernoulli trials: frequency and run quotas. *Statistics and Probability Letters* **9**, 5–11.

Erdös, P. and Rényi, A. (1970). On a new law of large numbers. *Journal d'Analyse Mathématique* **23**, 103–111.

Erdös, P. and Révész, P. (1975). On the length of the longest head-run. *Topics in Information Theory*, Colloquia Mathematica Societatis János Bolyai, 16 (eds. I. Csiszar and P. Elias; Keszthely, Hungary), North-Holland, Amsterdam, 219–228.

Ewan, W. D. and Kemp, K. W. (1960). Sampling inspection of continuous processes with no autocorrelation between successive results. *Biometrika* **47**, 363–380.

Feller, W. (1968). *An Introduction to Probability Theory and Its Applications* (Vol. I, 3rd ed.), Wiley, New York.

Fu, J. C. (1985). Reliability of consecutive-k-out-of-n:F system. *IEEE Transactions on Reliability* **34**, 127–130.

Fu, J. C. (1986). Reliability of consecutive-k-out-of-n:F systems with $(k-1)$ step Markov dependence. *IEEE Transactions on Reliability* **35**, 602–606.

Fu, J. C. (1995). Exact and limiting distributions of the number of successions in a random permutation. *Annals of the Institute of Statistical Mathematics* **47**, 435–446.

Fu, J. C. (1996). Distribution theory of runs and patterns associated with a sequence of multi-state trials. *Statistica Sinica* **6**, 957–974.

Fu, J. C. (2001). Distribution of scan statistics for a sequence of bi-state trials.

Journal of Applied Probability **38**, 1–9.

Fu, J. C. and Chang, Y. M. (2002). On probability generating functions for waiting time distributions of compound patterns in a sequence of multistate trials. *Journal of Applied Probability* **39**, 70–80.

Fu, J. C. and Hu, B. (1987). On reliability of a large consecutive-k-out-of-n:F system with $k-1$ step Markov dependence. *IEEE Transactions on Reliability* **36**, 75–77.

Fu, J. C. and Koutras, M. V. (1994). Distribution theory of runs: a Markov chain approach. *Journal of the American Statistical Association* **89**, 1050–1058.

Fu, J. C. and Lou, W. Y. W. (1991). On reliabilities of certain large linearly connected engineering systems. *Statistics and Probability Letters* **12**, 291–296.

Fu, J. C. and Lou, W. Y. W. (2000a). On the exact distribution of SECON and its application. *Statistica Sinica* **10**, 999–1010.

Fu, J. C. and Lou, W. Y. W. (2000b). Joint distribution of rises and falls. *Annals of the Institute of Statistical Mathematics* **52**, 415–425.

Fu, J. C., Lou, W. Y. W., Bai, Z. D. and Li, G. (2002). The exact and limiting distributions for the number of successes in success runs within a sequence of Markov-dependent two-state trials. *Annals of the Institute of Statistical Mathematics* **54**, 719–730.

Fu, J. C., Lou, W. Y. W. and Chen, S. C. (1999). On the probability of pattern matching in nonaligned DNA sequences: a finite Markov chain imbedding approach. *Scan Statistics and Applications* (eds. J. Glaz and N. Balakrishnan), Birkhäuser, Boston, 287–302.

Fu, J. C., Lou, W. Y. W. and Wang, Y. J. (1999). On the exact distributions of Eulerian and Simon Newcomb numbers associated with random permutations. *Statistics and Probability Letters* **42**, 115–125.

Fu, J. C., Shmueli, G. and Chang, Y. M. (2002). A unified Markov chain approach for computing the run length distribution for control charts with simple or compound rules. Technical Report, Department of Statistics, University of Manitoba.

Fu, J. C., Spring, F. A. and Xie, H. (2002). On the average run lengths of quality control schemes using a Markov chain approach. *Statistics and Probability Letters* **56**, 369–380.

Glaz, J. (1989). Approximations and bounds for the distribution of the scan statistic. *Journal of the American Statistical Association* **84**, 560–566.

Glaz, J. (1992). Approximations for tail probabilities and moments of the scan statistic. *Computational Statistics and Data Analysis* **14**, 213–227.

Glaz, J., Naus, J. I. and Wallenstein, S. (2001). *Scan Statistics*, Springer-Verlag, New York.

Godbole, A. P. (1990). Specific formulae for some success run distributions. *Statistics and Probability Letters* **10**, 119–124.

Godbole, A. P. (1991). Poisson approximations for runs and patterns of rare events. *Advances in Applied Probability* **23**, 851–865.

Goncharov, V. L. (1944). On the field of combinatory analysis. *Isvestija Akad. Nauk. SSSR. Ser. Math.* **8**, 3–48 (in Russian); English translation: *Translations of the AMS Ser. Math.* **19** (1962), 1–46.

Goodman, L. A. (1958). Simplified runs tests and likelihood ratio tests for Markoff chains. *Biometrika* **45**, 181–197.

Han, Q. and Aki, S. (1998). Formulae and recursions for the joint distributions of success runs of several lengths in a two-state Markov chain. *Statistics and Probability Letters* **40**, 203–214.

Han, Q. and Aki, S. (2000a). Sooner and later waiting time problems based on a dependent sequence. *Annals of the Institute of Statistical Mathematics* **52**, 407–414.

Han, Q. and Aki, S. (2000b). Waiting time problems in a two-state Markov chain. *Annals of the Institute of Statistical Mathematics* **52**, 778–789.

Hirano, K. (1986). Some properties of the distributions of order k. *Fibonacci Numbers and Their Applications* (eds. A. N. Philippou, G. E. Bergum, and A. F. Horadam), Reidel, Dordrecht, 43–53.

Hirano, K. and Aki, S. (1987). Properties of the extended distributions of order k. *Statistics and Probability Letters* **6**, 67–69.

Hirano, K. and Aki, S. (1993). One number of occurrences of success runs of specified length in a two-state Markov chain. *Statistica Sinica* **3**, 313–320.

Huntington, R. J. and Naus, J. I. (1975). A simpler expression for kth nearest neighbor coincidence probabilities. *Annals of Probability* **3**, 894–896.

Hwang, F. K. (1982). Fast solutions for consecutive-k-out-of-n:F system. *IEEE Transactions on Reliability* **31**, 447–448.

Hwang, F. K. (1986). Simplified reliabilities for consecutive-k-out-of-n systems. *SIAM Journal on Algebraic and Discrete Methods* **7**, 258–264.

Jackson, D. M. and Reilly, J. W. (1976). Permutations with a prescribed number of p-runs. *Ars Combinatoria* **1**, 297–305.

Johnson, B. C. (2001). Distribution of increasing l-sequences in a random permutation. *Methodology and Computing in Applied Probability* **3**, 35–49.

Johnson, B. C. (2002). The distribution of increasing 2-sequences in random permutations of arbitrary multi-sets. *Statistics and Probability Letters* **59**, 67–74.

Johnson, B. C and Fu, J. C. (2000). The distribution of increasing l-sequences in random permutations: A Markov chain approach. *Statistics and Probability Letters* **49**, 337–344.

Kaplansky, I. (1944). Symbolic solution of certain problems in permutations. *Bulletin of the American Mathematical Society* **50**, 906–914.

Karlin, S. and McGregor, J. (1959). Coincident probabilities. *Pacific Journal of Mathematics* **9**, 1141–1164.

Kontoleon, J. M. (1980). Reliability determination of a r-successive-out-of-n:F system. *IEEE Transactions on Reliability* **29**, 437.

Kossow, A. and Preuss, W. (1989). Reliability of consecutive-k-out-of-n:F system with nonidentical component reliabilities. *IEEE Transaction on Reliability*

38, 229–233.

Koutras, M. V. (1996a). On a Markov chain approach for the study of reliability structures. *Journal of Applied Probability* **33**, 357–367.

Koutras, M. V. (1996b). On a waiting time distribution in a sequence of Bernoulli trials. *Annals of the Institute of Statistical Mathematics* **48**, 789–806.

Koutras, M. V. (1997a). Waiting time distributions associated with runs of fixed length in two-state Markov chains. *Annals of the Institute of Statistical Mathematics* **49**, 123–139.

Koutras, M. V. (1997b). Waiting times and number of appearances of events in a sequence of discrete random variables. *Advances in Combinatorial Methods and Applications to Probability and Statistics* (ed. N. Balakrishnan), Birkhäuser, Boston, 363–384.

Koutras, M. V. (2003). Applications of Markov chains to the distribution theory of runs and patterns. *Handbook of Statistics 21: Stochastic Processes, Modeling and Simulation* (eds. D. N. Shanbhag and C. R. Rao), Elsevier, Amsterdam, in press.

Koutras, M. V. and Alexandrou, V. (1995). Runs, scans and urn model distributions: a unified Markov chain approach. *Annals of the Institute of Statistical Mathematics* **47**, 743–766.

Koutras, M. V. and Alexandrou, V. (1997a). Non-parametric randomness tests based on success runs of fixed length. *Statistics and Probability Letters* **32**, 393–404.

Koutras, M. V. and Alexandrou, V. (1997b). Sooner waiting time problems in a sequence of trinary trials. *Journal of Applied Probability* **34**, 593–609.

Koutras, M. V. and Papastavridis, S. G. (1993). Application of the Stein-Chen method for bounds and limit theorems in the reliability of coherent structures. *Naval Research Logistics* **40**, 617–631.

Ling, K. D. (1992). A generalization of the sooner and later waiting time problems for Bernoulli trials: frequency quota. *Statistics and Probability Letters* **14**, 401–405.

Ling, K. D and Low, T. Y. (1993). On the soonest and the latest waiting time distributions: succession quotas. *Communications in Statistics – Theory and Methods* **22**, 2207–2221.

Lou, W. Y. W. (1996). On runs and longest run tests: method of finite Markov chain imbedding. *Journal of the American Statistical Association* **91**, 1595–1601.

Lou, W. Y. W. (1997). An application of the method of finite Markov chain imbedding to runs tests. *Statistics and Probability Letters* **31**, 155–161.

Lou, W. Y. W. (2000). The exact distribution of the continuity of care measure *NOP*. *Statistics and Probability Letters* **48**, 361–368.

Lou, W. Y. W. (2001). The distribution of the usual provider continuity index under Markov dependence. *Statistics and Probability Letters* **54**, 269–276.

Lou, W. Y. W. (2003). The exact distribution of the K-tuple statistic for sequence homology. *Statistics and Probability Letters* **1**, 51–59.

Lucas, J. M. and Crosier, R. B. (1982). Fast initial response for CUSUM quality control schemes: Give your CUSUM a head start. *Technometrics* **24**, 199–205.

MacMahon, P. A. (1915). *Combinatory Analysis*, Cambridge University Press, London.

Mohanty, S. G. (1994). Success runs of length k in Markov dependent trials. *Annals of the Institute of Statistical Mathematics* **46**, 777–796.

Montgomery, D. C. (2001). *Introduction to Statistical Quality Control* (4th ed.). Wiley, New York.

Mood, A. M. (1940). The distribution theory of runs. *Annals of Mathematical Statistics* **11**, 367–392.

Mosteller, F. (1941). Note on an application of runs to quality control charts. *Annals of Mathematical Statistics* **12**, 228–232.

Muselli, M. (2000). Useful inequalities for the longest run distribution. *Statistics and Probability Letters* **46**, 239–249.

Nagaev, S. V. (1957). Some limit theorems for stationary Markov chains. *Theory of Probability and its Applications* **2**, 378–406.

Naus, J. I. (1965). The distribution of the size of the maximum cluster of points on a line. *Journal of the American Statistical Association* **60**, 532–538.

Naus, J. I. (1974). Probabilities for a generalized birthday problem. *Journal of the American Statistical Association* **69**, 810–815

Naus, J. I. (1982). Approximations for distributions of scan statistics. *Journal of the American Statistical Association* **77**, 177–183.

Nishimura, K. and Sibuya, M. (1997). Extended Stirling family of discrete probability distributions. *Communications in Statistics – Theory and Methods* **26**, 1727–1744.

Papastavridis, S. G. (1988). A Weibull limit for the reliability of a consecutive k-within-m-out-of-n system. *Advances in Applied Probability* **20**, 690–692.

Papastavridis, S. G. and Koutras, M. V. (1993). Bounds for reliability of consecutive k-within-m-out-of-n:F systems. *IEEE Transactions on Reliability* **42**, 156–160.

Philippou, A. N. (1986). Distributions and Fibonacci polynomials of order k, longest runs, and reliability of consecutive-k-out-of-n:F systems. *Fibonacci Numbers and Their Applications* (eds. A. N. Philippou, G. E. Bergum and A. F. Horadam), Reidel, Dordrecht, 203–227.

Philippou, A. N., Georghiou, C. and Philippou, G. N. (1983). A generalized geometric distribution and some of its properties. *Statistics and Probability Letters* **1**, 171–175.

Philippou, A. N. and Makri, F. S. (1986). Success runs and longest runs. *Statistics and Probability Letters* **4**, 211–215.

Pyke, R. (1961). Markov renewal processes: definitions and preliminary properties. *Annals of Mathematical Statistics* **32**, 1231–1242.

Reilly, J. W. and Tanny, S. M. (1979). Counting successions in permutations. *Studies in Applied Mathematics* **61**, 73–81.

Rényi, A (1970). *Probability Theory*, American Elsevier Publishing Company Inc., New York.

Riordan, J. (1958). *An Introduction to Combinatorial Analysis*, Wiley, New York.

Roselle, D. P. (1968). Permutations by number of rises and successions. *Proceedings of the American Mathematical Society* **19**, 8–16.

Ross, S. M. (2000). *Introduction to Probability Models* (7th ed.), Academic Press, San Diego.

Rubin, G., McCulloch, C. E. and Shapiro, M. A. (1990). Multinomial runs tests to detect clustering in constrained free recall. *Journal of the American Statistical Association* **85**, 315–320.

Saperstein, B. (1972). The generalized birthday problem. *Journal of the American Statistical Association* **67**, 425–428.

Schilling, M. F. (1990). The longest run of heads. *The College Mathematics Journal* **21**, 196–207.

Seneta, E. (1981). *Non-negative Matrices and Markov Chains* (2nd ed.), Springer-Verlag, New York.

Sheng, K. N. and Naus, J. I. (1994). Pattern matching between two non-aligned random sequences. *Bulletin of Mathematical Biology* **56**, 1143–1162.

Steinwachs, D. M. (1979). Measuring provider continuity in ambulatory care. *Medical Care* **17**, 551–565.

Swed, F. S. and Eisenhart, C. (1943). Tables for testing randomness of grouping in a sequence of alternatives. *Annals of Mathematical Statistics* **14**, 66–87.

Tanny, S. (1973). A probabilistic interpretation of Eulerian numbers. *Duke Mathematical Journal* **40**, 717–722.

Tanny, S. M. (1976). Permutations and successions. *Journal of Combinatorial Theory, Series A* **21**, 196–202.

Vaggelatou, E. (2003). On the length of the longest run in a multi-state Markov chain. *Statistics and Probability Letters* **62**, 211–221.

Uchida, M. and Aki, S. (1995). Sooner and later waiting time problems in a two-state Markov chain. *Annals of the Institute of Statistical Mathematics* **47**, 415–433

Wald, A. and Wolfowitz, J. (1940). On a test whether two samples are from the same population. *Annals of Mathematical Statistics* **11**, 147–162.

Wigle, D. T. (1982). Prevalence of selected chronic diseases in Canada, 1978–1979. *Chronic Disease in Canada* **3**, 9.

Wishart, J. and Hirshfeld, H. O. (1936). A theorem concerning the distribution of joins between line segments. *Journal of the London Mathematical Society* **11**, 227–235.

Wolfowitz, J. (1943). On the theory of runs with some applications to quality control. *Annals of Mathematical Statistics* **14**, 280–288.

Worpitzky, J. (1883). Studien über die Bernoullischen und Eulerschen Zahlen. *Journal für die reine und angewandte Mathematik* **94**, 203–232.

Index